Lecture Notes in Mathematics

A collection of informal reports and seminars
Edited by A. Dold, Heidelberg and B. Eckmann, Zürich

173

John N. Mordeson
Creighton University, Omaha, NB/USA
Bernard Vinograde
Iowa State University, Ames, IA/USA

Structure of Arbitrary Purely Inseparable Extension Fields

Springer-Verlag
Berlin · Heidelberg · New York 1970

ISBN 3-540-05295-X Springer-Verlag Berlin · Heidelberg · New York
ISBN 0-387-05295-X Springer-Verlag New York · Heidelberg · Berlin

Offsetdruck: Julius Beltz, Weinheim/Bergstr.

Preface

Starting with O. Teichmuller's basic concepts [56], G. Pickert developed an extensive theory of purely inseparable extensions, especially the finite degree case [43]. In these Notes we present an infinite degree theory, especially for the case without exponent. In addition to our own research, we include many relevant results from other sources which are acknowledged in the Reference Notes following each chapter. We stop short of the emerging Galois theory but have listed a number of references that may be consulted. It is assumed that the reader is acquainted with the elements of purely inseparable extensions such as appear in Jacobson [24].

Throughout these Notes L/K always denotes a field extension of a field K of characteristic $p \neq 0$. $|S|$ means the cardinality of a set S and $\subset$ means proper containment.

Contents

I. Generators

A. Relative p-bases. We collect here some old and some new facts about p-bases that will be used frequently in our work.

1.1. Definition. A relatively p-independent subset B of L/K is a subset B of L such that for all proper subsets B' of B, $K(L^p,B') \subset K(L^p,B)$. When K is perfect, B is the usual (absolutely) p-independent subset of L.

1.2. Definition. A relative p-base M of L/K is a subset M of L such that M is relatively p-independent in L/K and $L = K(L^p,M)$. When K is perfect, M is the usual p-base of L.

1.3. Definition. A minimal generating set M of L/K is a subset M of L such that $L = K(M)$ and for all proper subsets M' of M, $K(M') \subset L$. A subset M of L is called a minimal set over K if M is a minimal generating set of $K(M)/K$.

1.4. Proposition. Let L/K be purely inseparable and let M be a subset of L. Then M is a minimal generating set of L/K if and only if $L = K(M)$ and M is a relative p-base of L/K.

Proof. If M is a minimal generating set of L/K, then obviously $L = K(M)$ and thus $L = K(L^p,M)$. If M is not relatively p-independent in L/K, then there exists $m \in M$ such that $m \in K(L^p,M-m) = K(m^p,M-m)$. Thus m is both purely inseparable and separable algebraic over $K(M-m)$, whence $m \in K(M-m)$. However this contradicts the fact that M is a minimal set over K. Conversely if $L = K(M)$ and M is a relative p-base of L/K, then M is a minimal set over $K(L^p)$ whence a minimal set over K. q.e.d.

1.5. Definition. L/K is said to have an exponent (or, to be of bounded exponent) if and only if there exists a non-negative integer e such that $L^{p^e} \subseteq K$, where $L^{p^e} = \{a^{p^e} \mid a \in L\}$. The smallest such integer is called the exponent of L/K. A purely inseparable extension without exponent is also said to be of unbounded exponent. (Note that an extension with exponent must be purely inseparable.)

1.6. Corollary. Let L/K have exponent e and let M be a subset of L. Then M is a minimal generating set of L/K if and only if M is a relative p-base of L/K.

Proof. If M is a relative p-base of L/K, then $L = K(L^p, M)$ whence $L = K(L^{p^e}, M) = K(M)$. Hence the conclusion follows from Proposition 1.4. q.e.d.

1.7. Proposition. Let B and C be subsets of L/K and let e be a positive integer. Then

(a) B is relatively p-independent in L/K and C is a minimal generating set of $K(L^p)(B,C)/K(L^p)(B)$ if and only if $B \cup C$ is relatively p-independent in L/K and $B \cap C = \emptyset$.

(b) If $B \cup C$ is relatively p-independent in L/K and $B \cap C = \emptyset$, then $B \cap C^{p^{-e}}$ is relatively p-independent in $L(C^{p^{-e}})/K$.

(c) If $B \cup C$ is a relative p-base of L/K and $B \cap C = \emptyset$, then $B \cup C^{p^{-e}}$ is a relative p-base of $L(C^{p^{-e}})/K$.

Proof. (a) Suppose B is relatively p-independent in L/K and C is a minimal generating set of $K(L^p)(B,C)/K(L^p)(B)$. Then there does not exist $c \in C$ such that $c \in K(L^p)(B,C-c)$. If there exists $b \in B$ such that $b \in K(L^p)(B-b,C)$, then there exists $c \in C$ such that $b \notin K(L^p)(B-b,C-c)$. By the exchange property, $c \in K(L^p)(B,C-c)$, which is impossible. The converse is immediate. q.e.d.

(b) By induction, it suffices to prove the theorem for $e = 1$. If there exists $b \in B$ such that $b \in K((L(C^{p^{-1}}))^p)(B-b,C) = K(L^p)(C,B-b)$, then we contradict the relative p-independence of $B \cup C$ in L/K. Hence B is relatively p-independent in $L(C^{p^{-1}})/K$. We now apply part (a) by showing $C^{p^{-1}}$ is a minimal generating set of $K((L(C^{p^{-1}}))^p)(B,C^{p^{-1}})/K((L(C^{p^{-1}}))^p(B)$. Suppose there exists $c \in C$ such that $c^{p^{-1}} \in K(L^p(C))(B,C^{p^{-1}} - c^{p^{-1}})$. Then $c \in K^p(L^{p^2}(C^p))(B^p,C-c) \subseteq K(L^p)(C-c)$ which again contradicts the relative p-independence of $B \cup C$ in L/K.

(c) Suppose $B \cup C$ is a relative p-base of L/K. Then $B \cup C^{p^{-e}}$ is relatively p-independent in $L(C^{p^{-e}})/K$ by part (b). Now $K((L(C^{p^{-e}}))^p)(B, C^{p^{-e}}) = K(L^p(C,B))(C^{p^{-e}}) = L(C^{p^{-e}})$.

q.e.d.

<u>1.8. Proposition</u>. Let L' be an intermediate field of a purely inseparable extension L/K. If L'/K has a minimal generating set and $[L : L'] < \infty$, then L/K has a minimal generating set and the cardinality of a minimal generating set of L'/K is not greater than that of a minimal generating set of L/K.

Proof. By induction, it suffices to prove the proposition for L/L' simple, say $L = L'(b)$. Let e denote the exponent of b over L'. If $b^{p^e} \notin K(L'^p)$, then $\{b^{p^e}\}$ is a relatively p-independent subset of L'. Let M be a minimal generating set of L'/K (whence a relative p-base of L'/K by Proposition 1.4). Then by the exchange property there exists $m \in M$ such that $(M - m) \cup \{b^{p^e}\}$ is a relative p-base of L'/K. By Proposition 1.7, $(M - m) \cup \{b\}$ is a relative p-base of L/K. If t is the exponent of b over K, then $L = K(L^{p^t})(M - m, b) = K(m^{p^t})(M - m, b)$. Therefore, $m \in K(m^{p^t})(M - m, b)$, whence $L = K(M - m, b)$. Hence $(M - m) \cup \{b\}$ is a minimal generating set of L/K. Clearly $|(M - m) \cup \{b\}| = |M|$. On the other hand, suppose $b^{p^e} \in K(L'^p)$. Since $L = K(M, b)$, it suffices to show that $M \cup \{b\}$ is relatively p-independent in L/K. If $b \in K(L^p)(M)$, then $b \in K(L'^p)(b^p,M) = L'(b^p)$ which is impossible. If there exists $m \in M$ such that $m \in K(L^p)(M-m,b)$, then since $m \notin K(L^p)(M-m)$, we have by the exchange property that $b \in K(L^p)(M)$. Hence $b \in K(M^p,b^p)(M) = L'(b^p)$ which is impossible. Therefore $M \cup \{b\}$ is a minimal generating set of L/K and clearly $|M \cup \{b\}| \geq |M|$. q.e.d.

1.9. Definition. Let L' be an intermediate field of L/K. We say that L' is distinguished in L/K or, that L/L' preserves relative p-independence with respect to K, if and only if every relatively p-independent subset of L'/K is also a relatively p-independent subset of L/K. When K is perfect and L' is distinguished in L/K, we say that L/L' preserves p-independence.

1.10. Proposition. Let L' be an intermediate field of L/K. L' is distinguished in L/K if and only if there exists a relative p-base of L'/K which is relatively p-independent in L/K.

Proof. Suppose there exists a relative p-base M of L'/K which is relatively p-independent in L/K. If there exists a set X in L' which is relatively p-independent in L'/K, but is not relatively p-independent in L/K, then there exists a finite subset $X' \cup \{x\}$ of X such that $x \notin X'$ and $x \in K(L^p)(X')$. Let $M' \subseteq M$ be a finite set such that

$X' \cup \{x\} \subseteq K(L'^p)(M')$. Then there exists a subset M'' of M' such that $K(L'^p)(M') = K(L'^p)(X',x,M'')$ and $X' \cup \{x\} \cup M''$ is relatively p-independent in L'/K and M' and $X' \cup \{x\} \cup M''$ have the same number of elements, say n. Thus $K(L^p)(M') = K(L^p)(X',x,M'')$. Hence $[K(L^p)(M') : K(L^p)] = p^n > p^{n-1} \geq [K(L^p)(X',x,M'') : K(L^p)]$ which is impossible since $K(L^p)(M') = K(L^p)(X',x,M'')$. The converse is trivial. q.e.d.

1.11. Proposition. Let L' be a distinguished intermediate field of L/K. Let M and M' be subsets of L such that $M \cap M' = \emptyset$ and $M' \subseteq L'$. Then any pair of the following conditions implies the third condition.

(a) M is a relative p-base of L/L'.

(b) M' is a relative p-base of L'/K.

(c) $M \cup M'$ is a relative p-base of L/K.

Proof. (a) and (b) imply (c): If there exists $m \in M$ such that $m \in K(L^p)(M',M-m)$, then $m \in K(L^p)(L')(M-m) = L'(L^p)(M-m)$, a contradiction. If there exists $m' \in M'$ such that $m' \in K(L^p)(M'-m',M)$, then by condition (a) there exists $m \in M$ such that $m' \notin K(L^p)(M'-m',M-m)$ since L' is distinguished in L/K. Hence $m \in K(L^p)(M',M-m) \subseteq L'(L^p)(M-m)$ which contradicts condition (a). Thus $M \cup M'$ is relatively p-independent in L/K. Finally $K(L^p)(M,M') \supseteq L'$ whence $K(L^p)(M,M') = L$.

(b) and (c) imply (a): If there exists $m \in M$ such that $m \in L'(L^p)(M-m)$, then $m \in K(L'^p)(M')(L^p)(M-m) = K(L^p)(M',M-m)$ which contradicts condition (c). Also $L = K(L^p)(M',M)$ $= L'(L^p)(M)$.

(a) and (c) imply (b): By condition (c), M' is relatively p-independent in L'/K. Suppose $L' \supset K(L'^p)(M')$. Then there exists an element a in L' such that $M' \cup \{a\}$ is relatively p-independent in L'/K whence in L/K by hypothesis. Thus $a \notin K(L^p)(M')$, but $a \in K(L^p)(M',M)$. Hence there exists $m \in M$ such that $m \in K(L^p)(M',a,M-m) \subseteq L'(L^p)(M-m)$ which contradicts condition (a). q.e.d.

<u>1.12. Proposition</u>. Let L' be an intermediate field of a purely inseparable extension L/K such that L'/K has a minimal generating set. Then every minimal generating set of L'/K can be extended to a minimal generating set of L/K if and only if L' is distinguished in L/K and L/L' has a minimal generating set.

Proof. Suppose every minimal generating set M' of L'/K can be extended to a minimal generating set of L/K, say $M' \cup M$. Then M' and $M' \cup M$ are relative p-bases of L'/K and L/K, respectively. Hence L' is distinguished in L/K by Proposition 1.10. By Proposition 1.11, M is a relative p-base of L/L'. Since $L = K(M',M)$, $L = L'(M)$. Thus M is a minimal generating set of L/L'. Conversely, suppose L' is distinguished in L/K and L/L' has a minimal generating set M. Then M is a relative p-base of L/L'. Let M' be a minimal generating set whence a relative p-base of L'/K. Then by Proposition 1.11, $M \cup M'$ is a relative p-base of L/K. Thus $M \cup M'$ must be a minimal generating set of L/K since it's a generating set of L/K. q.e.d.

1.13. Proposition. Let L' be an intermediate field of L/K.

(a) If L/L' is separable, then L' is distinguished in L/K.

(b) Let L/L' be algebraic and let K be perfect. If L' is distinguished in L/K, then L/L' is separable algebraic.

Proof. (a) Suppose L/L' is separable. Then L' and L^p are linearly disjoint over L'^p. Thus every p-base of L' is p-independent in L. Let M be a relative p-base of L'/K and let G be a subset of K such that $L'^p(G) = K(L'^p)$ and $G \cup M$ is a p-base of L'. If there exists $m \in M$ such that $m \in K(L^p)(M-m)$, then $m \in K(L'^p)(L^p, M-m) \subseteq L^p(G, M-m)$. However this contradicts the fact that every p-base of L' is p-independent in L.

(b) Suppose L' is distinguished in L/K. Since K is perfect, this means L/L' preserves p-independence. Let S denote the maximal separable intermediate field of L/L'. Suppose $L \supset S$. Then there exists $c \in L$, $c \notin S$ such that

$c^p \in S$. Any p-base B of L' is p-independent in S by part (a). Now $S^p(B) \supseteq L'^p(B) \supseteq L'$ whence S is both separable and purely inseparable over $S^p(B)$. Thus $S = S^p(B)$. Hence $c^p \in S^p(B)$. By the exchange property, there exists $b \in B$ such that $b \in S^p(B - b, c^p) \subseteq L^p(B - b)$. Hence B is not p-independent in L which contradicts the hypothesis that L/L' preserves p-independence. Therefore $L = S$. q.e.d.

B. Extensions of type R. A relative p-base is not necessarily a minimal generating set. This difference can occur only when L/K has no exponent.

1.14. Definition. L/K is said to be of type R if and only if every relative p-base of L/K is a generating set (hence a minimal generating set) of L/K.

1.15. Proposition. (a) L/K is of type R if and only if $L \neq L'(L^p)$ for every intermediate field L' of L/K such that $L' \neq L$.

(b) If L/K is of type R, then L/L' is of type R for every intermediate field L' of L/K.

Proof. (a) If $L = L'(L^p)$ where L' is an intermediate field of L/K such that $L' \neq L$, then L' contains a relative p-base M of L/K. Thus $L \supset L' \supseteq K(M)$. Conversely, if there exists a relative p-base M of L/K such that $L \supset K(M)$, then $L = L'(L^p)$, where $L' = K(M)$.

(b) $L \neq L''(L^p)$ for every intermediate field L'' of L/L' such that $L'' \neq L$ since $L'' \supseteq K$ and L/K is of type R.

q.e.d.

1.16. Lemma. Let L/K be purely inseparable and let $M = \{m_1, m_2, \ldots\}$ be a relative p-base of L/K. If there exist positive integers $e_1, e_2, \ldots$ such that $e_i < e_{i+1}$, $i = 1, 2, \ldots$, and such that $\{m_i^{p^{e_i - e_1}}\}_{i=1}^{\infty}$ is a minimal set over K, then L/K is not of type R.

Proof. Suppose L/K is of type R. Now, it is readily verified that $M' = \{m_i m_{i+1}^{p^{e_{i+1} - e_i}}\}_{i=1}^{\infty}$ is a relative p-base of L/K whence $L = K(M')$. Therefore there exists a positive integer r such that $m_1 \in L_r = K(m_1 m_2^{p^{e_2 - e_1}}, m_2 m_3^{p^{e_3 - e_2}}, \ldots, m_r m_{r+1}^{p^{e_{r+1} - e_r}})$. Hence $M_r = \{m_i^{p^{e_i - e_1}} \mid i = 1, \ldots, r+1\} \subseteq L_r$. Set $L_{r+1} = K(M_r)$. By our hypothesis M_r is a minimal set over K. Clearly, $\{m_1 m_2^{p^{e_2 - e_1}}, m_2 m_3^{p^{e_3 - e_2}}, \ldots, m_r m_{r+1}^{p^{e_{r+1} - e_r}}\}$ is a minimal generating set of L_r/K. Thus L_r/K is minimally generated by r elements while the intermediate field L_{r+1} is minimally generated by $r+1$ elements over K. However this is impossible by Proposition 1.8. Hence L/K is not of type R. q.e.d.

<u>1.17. Proposition</u>. Let L/K be purely inseparable and set $D(L) = \{L' \mid L'$ is distinguished in $L/K\}$. L/K has an exponent if and only if every element of $D(L)$ is of type R.

Proof. Suppose every element of $D(L)$ is of type R and that L/K has unbounded exponent. Let M be a relative p-base of L/K. Then $L = K(M)$ because $L \in D(L)$. Since L/K has unbounded exponent, M is an infinite set and there exists a sequence $\{m_i\}_{i=1}^{\infty}$ such that $m_i \in M$, $e_i < e_{i+1}$, $i = 1,2,\ldots,$ where e_i is the exponent of m_i over $K(m_1,\ldots,m_{i-1})$ and $K(m_0)$ means K. The strict inequality can be achieved because L/K would have bounded exponent if $L/K(m_1,\ldots,m_i)$ has bounded exponent for any finite subset $\{m_1,\ldots,m_i\}$ of M. Now consider the sequence $\{m_i^{p^{e_i-e_1}}\}_{i=1}^{\infty}$. Suppose there exists a subsequence $\{m_{i_j}\}_{j=1}^{\infty}$ of $\{m_i\}_{i=1}^{\infty}$ with the following properties: If we let $n_j = m_{i_j}$ and $f_j = e_{i_j}$, $j = 1,2,\ldots$ (with $n_1 = m_1$ for convenience and without loss of generality), then $\{n_j^{p^{f_j-f_1}}\}_{j=1}^{\infty}$ is a minimal set over K. Let $N' = \{n_j\}_{j=1}^{\infty}$. Then $K(N') \in D(L)$ whence $K(N')$ is of type R. However $\{n_j\}_{j=1}^{\infty}$ is a sequence satisfying the conditions of Lemma 1.16 so that $K(N')$ is not of type R. Thus the subsequence $\{n_j\}_{j=1}^{\infty}$ cannot exist. Therefore the intermediate field $L' = K(\{m_i^{p^{e_i-e_1}}\}_{i=1}^{\infty})$ has the property that every relative p-base of L'/K is finite because a relative p-base of L'/K containing m_1 can be chosen from the generating set $\{m_i^{p^{e_i - e_1}}\}_{i=1}^{\infty}$ and a subsequence of the above type does not exist. Thus if M' is a relative p-base of L'/K, then there exists a positive integer $t = t(M')$ such that $M'^{p^t} \subseteq K$. Hence $K(L'^{p^t}) = K(L'^{p^{t+1}})$, so $K(L'^{p^t})$ is relatively perfect over K. Since L'/K has unbounded exponent, $K(L'^{p^t})/K$ has unbounded exponent. But since $K(L'^{p^t})$ is relatively perfect over K, $K(L'^{p^t}) \in D(L)$ with $\emptyset$ as its relative p-base. Thus, by hypothesis, $K(L'^{p^t}) = K(\emptyset) = K$, a contradiction. It is of course immediate that if L/K has bounded exponent then every element of $D(L)$ is of type R. q.e.d.

<u>1.18. Corollary</u>. Let L/K be purely inseparable and set $S(L) = \{L' \mid L'$ is an intermediate field of $L/K\}$. L/K has an exponent if and only if every element of $S(L)$ is of type R.

<u>1.19. Proposition</u>. Let L/K be purely inseparable. Suppose L' and L'' are intermediate fields of L/K such that $L = L'L''$ and L''/K has an exponent. If L/K is of type R, then both L'/K and L''/K are of type R.

Proof. The conclusion for L''/K has already been noted. If there exists a relative p-base M of L'/K such that $L' \supset K(M)$, then there exists a proper intermediate field L^* of L' such that $L' = L^*(L'^p)$ and L'/L^* has unbounded exponent, in fact, such a field is $L^* = K(M)$. Now $L'' = K(N)$, where N is a minimal generating set, so $N^{p^t} \subseteq K$ for some positive integer t. Set $L^{**} = L^*(N)$. Then $L^{**p^t} \subseteq L^*$, whence $L^{**} \neq L$, else L/L^* and thus L'/L^* would have bounded exponent. Now $L = L'L'' = L'(N) = L^*L'^p(N) = L^{**}K(L'^p) \subseteq L^{**}(L^p) \subseteq L$. That is, $L = L^{**}(L^p)$ for a proper L^{**}, which contradicts part (a) of Proposition 1.15. q.e.d.

1.20. Proposition. Let L/K be purely inseparable. Set $F_c(L) = \{L' \mid L'$ is an intermediate field of L/K such that $[L : L'] < \infty$, that is, L' is cofinite in $L/K\}$.

(a) Suppose $L \neq K(L^p)$. L/K is of type R if and only if L'/K is of type R for every $L' \in F_c(L)$ such that $L' \neq L$.

(b) L/K is of type R if and only if L'/K is of type R for every $L' \in F_c(L)$.

Proof. (a) Suppose L/K is of type R. If L/L' is finite, then $L = L'(n_1, \ldots, n_r)$ for some finite subset $\{n_1, \ldots, n_r\}$ of L. Hence $L = L'L''$, where $L'' = K(n_1, \ldots, n_r)$, so L'/K is of type R by Proposition 1.19. Conversely, suppose every proper intermediate field L' of L/K such that L/L' is finite, is of type R. If L/K is not of type R, then $L = K(M)(L^p)$ and $L \supset K(M)$, where M is a relative p-base of L/K and $M \neq \emptyset$ because $L \supset K(L^p)$. Let $L' = K(L^p)(M - m)$ for some $m \in M$. Now $K(L^p) = K(M^p)(L^{p^2})$, whence $L' = K(M^p)(M - m)(L^{p^2}) = K(m^p)(M - m)(L'^p)$. If

$L' = K(m^p)(M-m)$, then $L = L'(m) = K(M)$, a contradiction. Thus $L' \supset K(m^p)(M-m)$ and so we have that L'/K is not of type R which contradicts the hypothesis. q.e.d.

(b) Suppose L/K is of type R and $L \neq K$. Then $L \supset K(L^p)$. Thus, by part (a), it follows that for every $L' \in F_c(L)$, L'/K is of type R. The converse is immediate because $L \in F_c(L)$. q.e.d.

1.21. Definition. A subset M of a purely inseparable extension L/K is called a subbase of L/K if and only if $M \subseteq L - K$, $L = K(M)$, and for every finite subset $\{m_1, \ldots, m_r\}$ of M, $[K(m_1, \ldots, m_r) : K] = \prod_{i=1}^{r} [K(m_i) : K]$, that is, $K(m_1, \ldots, m_r)$ is a tensor product over K of the simple extensions $K(m_1), \ldots, K(m_r)$.

1.22. Proposition. Let L/K be purely inseparable and B a p-base of L. Set $C = \{b^{p^i} \mid b \in B,\ i$ is the exponent of b over $K\}$. Then the following conditions are equivalent:

(1) $B - K$ is a subbase of L/K.

(2) $K = K^p(C)$.

(3) C is a p-base of K.

(4) $L = K(B)$ and $K \subseteq L^{p^i}(C)$, $i = 1, 2, \ldots$.

Proof. (1) implies (2): Let $\{b_1, \ldots, b_r\}$ be a finite subset of $B - K$ and let e_i be the exponent of b_i over K, $i = 1, \ldots, r$. Then, since $K^p(C) \subseteq K$,

$p^{e_1}\cdots p^{e_r} \geq [K^p(C)(b_1,\ldots,b_r) : K^p(C)] \geq [K(b_1,\ldots,b_r) : K] =$ $p^{e_1}\cdots p^{e_r}$. Therefore, K and $K^p(B)$ are linearly disjoint over $K^p(C)$, whence $K \cap K^p(B) = K^p(C)$. Since $L = K(B)$, $K \subseteq L = L^p(B) = K^p(B)$. Thus $K = K^p(C)$.

(2) implies (3): Since $K = K^p(C)$, there exists a subset C' of C such that C' is a p-base of K. Let $B' = \{b \mid b \in B, b^{p^i} \in C'\}$. Then $(K(B'))^p(B') =$ $K^p(B') = K^p(C')(B') = K(B')$. Hence B' is a p-base of $K(B')$. Thus $L/K(B')$ preserves p-independence, whence $L = K(B')$ by Proposition 1.13 and the pure inseparability of L/K. Thus $B = B'$ whence $C = C'$.

(3) implies (4): Since C is a p-base of K, we have $K = K^{p^i}(C)$, $i = 1,2,\ldots$. Therefore $K \subseteq L^{p^i}(C)$, $i = 1,2,\ldots$. Now $K = K^p(C)$ implies $K \subseteq K^p(B)$. Hence $(K(B))^p(B) = K(B)$ and thus B is a p-base of $K(B)$. Thus $L/K(B)$ preserves p-independence, from which it follows that $L = K(B)$.

(4) implies (1): Let $B_1 \cup \ldots \cup B_r$ be any finite subset of $B - K$ where every element of B_i has exponent i over K, $i = 1,\ldots,r$. Suppose that B_i has $s_i \geq 0$ elements $(i = 1,\ldots,r)$. Since C consists of p^e-th powers of elements of the p-base B of L, it follows that

(*) $[L^{p^{i+1}}(C)(B_{i+1}^{p^i},\ldots,B_r^{p^i}) : L^{p^{i+1}}(C)] = p^{s_{i+1}}\cdots p^{s_r}$. Now $[K(B_1,\ldots,B_r) : K] = [K(B_1,\ldots,B_r) : K(B_1^p,\ldots,B_r^p)] \cdot [K(B_1^p,\ldots,B_r^p) :$

$K(B_2^{p^2},\dots,B_r^{p^2}]\cdot \dots \cdot[K(B_r^{p^{r-1}}):K]$. Since $L^{p^{i+1}}(C) \supseteq K(B_{i+1}^{p^{i+1}},\dots,B_r^{p^{i+1}})$, we have $[K(B_{i+1}^{p^i},\dots,B_r^{p^i}) : K(B_{i+1}^{p^{i+1}},\dots,B_r^{p^{i+1}})] = p^{s_{i+1}}\dots p^{s_r}$, otherwise we contradict equation (*). Thus $[K(B_1,\dots,B_r) : K] = p^{s_1}\dots p^{s_r}p^{s_2}\dots p^{s_r} = p^{s_1}p^{2s_2}\dots p^{rs_r}$. Hence $B - K$ is a subbase of L/K. q.e.d.

<u>1.23. Proposition</u>. Let L/K be an extension and B a p-base of L. Set $C = \{b^{p^i} \mid b \in B,\ i$ is the exponent of b over K if b is purely inseparable over K and $i = 0$ otherwise$\}$. Each statement in the following list implies the succeeding one.

(1) L/K is purely inseparable and has a subbase.

(2) K and L^{p^i} are linearly disjoint, $i = 1,2,\dots$.

(3) There exists a p-base B of L such that $K \subseteq L^{p^i}(C)$, $i = 1,2,\dots$.

(4) If L/K is purely inseparable and of unbounded exponent, then L/K is not of type R.

Proof. (1) implies (2): Let M be a subbase of L/K. Then for all $i = 1,2,\dots$, $M = M_i' \cup M_i$, where every element of M_i' has exponent at most i over K and every element of M_i has exponent greater than i over K. Since $L = K(M_i', M_i)$, $L^{p^i} = K^{p^i}(M_i'^{p^i}, M_i^{p^i}) = (L^{p^i} \cap K)(M_i^{p^i})$, $i = 1,2,\dots$. For any finite subset $\{b_1,\dots,b_r\} \subseteq M_i$, let e_j denote the exponent of b_j over K, $j = 1,\dots,r$. Then

$p^{e_1-i} \ldots p^{e_r-i} \geq [(L^{p^i} \cap K)(b_1^{p^i}, \ldots, b_r^{p^i}) : L^{p^i} \cap K]$

$\geq [K(b_1^{p^i}, \ldots, b_r^{p^i}) : K] = p^{e_1-i} \ldots p^{e_r-i}$. Thus L^{p^i} and K are linearly disjoint over $L^{p^i} \cap K$, $i = 1,2,\ldots$.

(2) implies (3): Since L^p and K are linearly disjoint over $L^p \cap K$, there exists a set C_o in K such that $K = (L^p \cap K)(C_o)$ and C_o is p-independent in L. Suppose that there exist sets C_j, $j = 0,\ldots,i-1$, such that $C_j \subseteq L^{p^j} \cap K$, $L^{p^{i-1}} \cap K = (L^{p^i} \cap K)(C_o^{p^{i-1}}, C_1^{p^{i-2}}, \ldots, C_{i-1})$ and $C_o^{p^{i-1}} \cup C_1^{p^{i-2}} \cup \ldots \cup C_{i-1}$ is p-independent in $L^{p^{i-1}}$. Then $C_o^{p^i} \cup C_1^{p^{i-1}} \cup \ldots \cup C_{i-1}^p \subseteq L^{p^i} \cap K$ and is p-independent in L^{p^i}. Since $L^{p^{i+1}}$ and K are linearly disjoint over $L^{p^{i+1}} \cap K$, $L^{p^{i+1}}$ and $L^{p^i} \cap K$ are linearly disjoint over $L^{p^{i+1}} \cap K$. Thus there exists $C_i \subseteq L^{p^i} \cap K$ such that $L^{p^i} \cap K = (L^{p^{i+1}} \cap K)(C_o^{p^i}, C_1^{p^{i-1}}, \ldots, C_i)$ and $C_o^{p^i} \cup \ldots \cup C_i$ is p-independent in L^{p^i}. Hence there exist sets C_i, $i = 0,1,\ldots,$ such that $C_i \subseteq L^{p^i} \cap K$, $L^{p^i} \cap K = (L^{p^{i+1}} \cap K)(C_o^{p^i}, C_1^{p^{i-1}}, \ldots, C_i)$ and $C_o^{p^i} \cup C_1^{p^{i-1}} \cup \ldots \cup C_i$ is p-independent in L^{p^i}, $i = 0,1,\ldots$. Thus $K = (L^{p^i} \cap K)(C_o, \ldots, C_{i-1})$ whence $K = (L^{p^i} \cap K)(C_o, C_1, \ldots)$, $i = 1,2,\ldots$. Furthermore, $\bigcup_{i=0}^{\infty} C_i^{p^{-i}}$ is p-independent in L.

Augment $\bigcup_{i=0}^{\infty} C_i^{p^{-i}}$ to a p-base B of L. Then for $C = \{b^{p^i} \mid i \in B,$ i is the exponent of b over K if b is purely inseparable over K and i = 0 otherwise$\}$,

$$(*) \qquad K = (L^{p^i} \cap K)(C^*),\ i = 1,2,\ldots,\ C^* = C \cap K,$$

since $C^* \supseteq \bigcup_{i=0}^{\infty} C_i$. Thus $K \subseteq L^{p^i}(C)$, $i = 1,2,\ldots$.

(3) implies (4): Let B be a p-base of L satisfying condition (3), where L/K is purely inseparable and of unbounded exponent. Suppose first that $L \supset K(B)$. Then since $L = K(L^p)(B)$, there exists a relative p-base M of L/K such that $M \subseteq B$. Hence $K(M) \subseteq K(B) \subset L$. Suppose that $L = K(B)$. Set $M = B - K$. Then by Proposition 1.22, M is a subbase of L/K. Hence there exist $m_1, m_2, \ldots \in M$ such that m_i has exponent e_i $(e_i < e_{i+1})$ over $K' = K(M')$, where $M' = M - \{m_1, m_2, \ldots\}$, $i = 1,2,\ldots$. Now it is readily verified that $\{m_1, m_2, \ldots\}$ is a relative p-base of L/K' satisfying the hypotheses of Lemma 1.16. Hence there exists a relative p-base M" of L/K' such that $L \supset K'(M")$. By Proposition 1.11 $M' \cup M"$ is a relative p-base of L/K and we have $L \supset K(M',M")$. Thus L/K is not of type R. q.e.d.

<u>1.24. Proposition</u>. Let L/K be purely inseparable and without exponent. Let M be a relative p-base of L/K. Each statement in the following list implies the succeeding one.

(1) M is a subbase of L/K.

(2) $\bigcap_{m \in M} K(M-m) = K$.

(3) $\bigcap_{m \in M} K(M-m) \not\supseteq L^{p^i}$, $i = 1,2,\ldots$.

(4) L/K is not of type R.

Proof. (1) implies (2): This is immediate from Definition 1.21.

(2) implies (3): If $\bigcap_{m \in M} K(M-m) \supseteq L^{p^e}$ for some positive integer e, then $L^{p^e} \subseteq K$ which is impossible.

(3) implies (4): If $L \supset K(M)$, then L/K is not of type R. Suppose $L = K(M)$. For all $m \in M$, let e_m denote the exponent of L over $K(M-m)$. Then $\{e_m \mid m \in M\}$ is unbounded else $\{e_m \mid m \in M\}$ has a least upper bound e, whence $L^{p^e} \subseteq \bigcap_{m \in M} K(M-m)$, contradicting condition (3). Thus there exists a subset $\{e_1, e_2, \ldots\} \subseteq \{e_m \mid m \in M\}$ such that $e_1 < e_2 < \ldots$. For m_i corresponding to e_i, $i = 1,2,\ldots$, the sequence $\{m_i\}_{i=1}^{\infty}$ is such that $m_i^{p^{e_i-e_1}} \notin K(M-m_i)$. Set $M' = M - \{m_1, m_2, \ldots\}$ and $K' = K(M')$. Now $m_i^{p^{e_i-e_1}} \notin K(M-m_i)$ implies $m_i^{p^{e_i-e_1}} \notin K'(\{m_j^{p^{e_j-e_1}}\}_{j=1}^{\infty} - m_i^{p^{e_i-e_1}})$. Thus $\{m_i^{p^{e_i-e_1}}\}_{i=1}^{\infty}$ is a minimal set over K'. That is, $\{m_1, m_2, \ldots\}$ is a set satisfying the conditions in Lemma 1.6 (with K replaced by K'). Hence, as in the proof of (3) implies (4) of Proposition 1.23, L/K is not of type R. q.e.d.

<u>1.25. Proposition</u>. Let L/K be purely inseparable. If $L = K(m_1, m_2, \ldots)$ where $m_i \in K(m_{i+1})$, $i = 1,2,\ldots,$ then the intermediate fields of L/K are K, $K(m_i^{p^{j_i}})$ and L, $0 \leq j_i < e_i$ (e_i the exponent of m_i over $K(m_{i-1})$), $i = 1,2,\ldots,$ where $K(m_0)$ means K.

Proof. Let e_i' denote the exponent of m_i over K, $i = 1,2,\ldots$. Because $K(m_s)$ is simple, the intermediate fields of $K(m_s)/K$ are $K(m_s^{p^{j_s}})$, $0 \leq j_s \leq e_s'$. If $0 < t < s$, then $K(m_t) \subset K(m_s)$, whence $K(m_t) = K(m_s^{p^{e_s'-e_t'}})$. Thus the intermediate fields of $K(m_s)/K$ are K, $K(m_i^{p^{j_i}})$, $0 \leq j_i < e_i$, $i = 1,\ldots,s$. Let K' be any intermediate field of L/K. If $[K' : K] < \infty$, then K'/K is finitely generated. Hence $K' \subseteq K(m_s)$ for some m_s since $L = \bigcup_{i=1}^{\infty} K(m_i)$. Thus $K' = K(m_i^{p^{j_i}})$ for some m_i by the preceding argument. If $[K' : K] = \infty$, then K' is the union over c of $K(c)$ for all $c \in K'$. Now $K(c) = K(m_{i_c}^{p^{j_{i_c}}}) \supset K(m_{i_c-1})$ for some m_{i_c} and for all $c \in K' - K$ by the previous argument. Since $[K' : K] = \infty$, i_c is an unbounded function of c. Thus $K' = L$. q.e.d.

<u>1.26. Example</u>. There exist purely inseparable extensions which are simultaneously of type R and without exponent.

(a) L/K is of unbounded exponent, the maximal perfect subfield of L is not contained in K, and L/K is of type R: Let P be a perfect field and $z, y, x_1, x_2, \ldots$ independent indeterminates over P. Let $K = P(z, y, x_1, x_2, \ldots)$ and $L = K(m_1, m_2, \ldots)$, where $m_i = z^{p^{-i-1}} x_i^{p^{-1}} + y^{p^{-1}}$, $i = 1, 2, \ldots$. Clearly L/K is of unbounded exponent. $P(z^{p^{-1}}, z^{p^{-2}}, \ldots)$ is the maximal perfect subfield and is not in K. By Proposition 1.15, L/K is of type R if we show that $L \neq L'(L^p)$ for any intermediate field L' of L/K such that $L' \neq L$. We postpone this proof.

(b) L/K is of unbounded exponent, the maximal perfect subfield of L is contained in K and L/K is of type R: Let P be a perfect field and $y, x_1, x_2, \ldots$ independent indeterminates over P. Let $K = P(y, x_1, x_2, \ldots)$ and $L = K(m_1, m_2, \ldots)$, where $m_1 = x_1^{p^{-2}}$ and $m_{i+1} = (m_i^p y + x_{i+1})^{p^{-2}}$, $i = 1, 2, \ldots$. Clearly L/K is of unbounded exponent. It follows that $L = P(y, m_1, m_2, \ldots)$ and that $\{y, m_1, m_2, \ldots\}$ is an algebraically independent set over P. That is, L/P is a pure transcendental extension. Thus P, $P \subseteq K$, is the maximal perfect subfield of L by the Corollary in [9, p.20]. By Proposition 1.15, it remains to be shown that $L \neq L'(L^p)$ for any proper intermediate field L' of L/K.

We prove simultaneously for the examples in parts (a) and (b) that such a field L' cannot exist. In both examples it follows that $K(L^p) = K(m_1^p, m_2^p, \ldots)$, $m_i^p \in K(m_{i+1}^p)$ and m_{i+1}^p

has exponent 1 over $K(m_i^p)$ for $i = 1,2,\ldots$. Hence by Proposition 1.25, the intermediate fields of $K(L^p)/K$ appear in a chain. Now suppose there exists an intermediate field L' of L/K such that $L = L'(L^p)$ and $L' \neq L$. Since $L' \not\supseteq K(L^p)$, $L' \cap K(L^p) = K(m_s^p)$ for some integer $s \geq 0$. We show L' and $K(L^p)$ are linearly disjoint over $K(m_s^p)$ by showing that for every proper intermediate field K' of $K(L^p)/K(m_s^p)$, L' and K' are linearly disjoint over $K(m_s^p)$. By Proposition 1.25, $K' = K(m_t^p)$ for $t \geq s$. Now m_t^p has exponent $t - s$ over $K(m_s^p) \subseteq L'$. If $((m_t^p)^{p^{t-s}})^{p^{-1}} \in L'$, then we contradict $L' \cap K(L^p) = K(m_s^p)$. Hence the irreducible polynomial of m_t^p over $K(m_s^p)$ remains irreducible over L'. Thus L' and $K(m_t^p)$ are linearly disjoint over $K(m_s^p)$, whence L' and $K(L^p)$ are linearly disjoint.

Since $L = L'(L^p)$, $m_{s+1} \in L'(L^p)$. Hence

$$m_{s+1} = \sum_{j=0}^{p^{t-s}-1} c_j'(m_t^p)^j \quad \text{for some integer } t, \text{ where } c_j' \in L'.$$

Now $t \geq s + 2$ since m_{s+1} has exponent 2 over $K(m_s^p)$ and L' has exponent 1 over $K(m_s^p)$. Thus $m_{s+1}^p = \sum_j c_j'^p (m_t^p)^{jp}$. By the division algorithm, $jp = p^{t-s}q_j + r_j$, $0 \leq r_j < p^{t-s}$. Hence

$$(*) \qquad m_{s+1}^p = \sum_j (c_j'^p (m_t^p)^{p^{t-s}q_j})(m_t^p)^{r_j}$$

and

$$c_j'^p (m_t^p)^{p^{t-s}q_j} = c_j^p \in L' \cap L^p.$$

Writing m_{s+1}^p in terms of m_t^p, we get for the example in part (a): $m_{s+1}^p = (m_t^p)^{p^{t-s-1}} k_o^p x_{s+1} - k_1$, where $k_o^p = x_t^{-p^{t-s-1}}$ and $k_1 = y^{p^{t-s-1}} x_t^{-p^{t-s-1}} x_{s+1} - y$. By equation (*), $(m_t^p)^{p^{t-s-1}} x_{s+1} = k_o^{-p} k_1 + k_o^{-p} \sum_j c_j^p (m_t^p)^{r_j}$. Hence, by the linear disjointness of L' and $K(L^p)$ over $K(m_s^p)$ and since $\{(m_t^p)^j \mid j = 0,\ldots,p^{t-s} - 1\}$ is linearly independent over $K(m_s^p)$, $x_{s+1} \in L' \cap L^p$. Thus $x_{s+1}^{p^{-1}} \in L$, a contradiction.

For the example in part (b), we get $m_{s+1}^p = (m_{s+2}^p)^p y^{-1} - x_{s+2} y^{-1} = \ldots = (m_t^p)^{p^{t-s-1}} k_o^p y^{-1} - k_1$ for suitable $k_o, k_1 \in K$. By an argument similar to that of the example in part (a), we obtain $y^{p^{-1}} \in L$, a contradiction.

If L/K is of type R and F/L is a finite degree purely inseparable field extension, then F/K is not necessarily of type R. For instance, if L/K is the example in part (a) above and $F = L(y^{p^{-1}})$, then $M = \{y^{p^{-1}}, x_1^{p^{-1}}, x_2^{p^{-1}}, \ldots\}$ is a relatively p-base of F/K such that $F \neq K(M)$. However the following proposition gives a criterion for an extension F/L to be of type R when L/K is of type R.

<u>1.27. Proposition</u>. Let F/L be an extension of an extension L/K. If L/K is of type R and $L = K(F^{p^e})$ for some positive integer e, then F/K is of type R.

Proof. Let N be a relative p-base of F/K. Then $L \supseteq K(L^p)(N^{p^e}) = K(F^{p^{e+1}})(N^{p^e}) = K(F^{p^e}) = L$. Hence N^{p^e} contains a relative p-base M of L. Thus $F = K(F^{p^e})(N) = L(N) = K(M)(N) = K(N)$. q.e.d.

1.28. Corollary. If L/K is an extension such that $K(L^{p^e})/K$ is of type R for some positive integer e, then L/K is of type R.

C. Special generating systems. We now derive and analyze some generating systems associated with the towers
$L \supseteq K(L^p) \supseteq K(L^{p^2}) \supseteq \ldots$ and $K \subseteq K^{p^{-1}} \cap L \subseteq K^{p^{-2}} \cap L \subseteq \ldots$.

1.29. Proposition. Let L/K be purely inseparable and let $M_0, M_1, \ldots$ be subsets of L such that $M_i \supseteq M_{i+1}$ and $M_i^{p^i}$ is a relative p-base of $K(L^{p^i})/K$, $i = 0,1,\ldots$.

A. The subsets $B_i = M_{i-1} - M_i$, $i = 1,2,\ldots$, satisfy:

(a) $\bigcup_{i=1}^{\infty} B_i = M_0$ and for $i = 1,2,\ldots$,

(1) $B_i^{p^i} \subseteq K(L^{p^{i+1}})(B_{i+1}^{p^i}, B_{i+2}^{p^i}, \ldots)$,

(2) For all $b \in B_i$, b has exponent i over $K(L^{p^{i+1}})(B_i - b, B_{i+1}, \ldots)$.

(b) $\bigcup_{j=i}^{\infty} B_j^{p^i}$ is p-independent in $K(L^{p^i})$, $i = 1,2,\ldots$.

(c) The cardinalities of B_i and M_{i-1}, $i = 1,2,\ldots$, are invariants of the extension.

B. Given a generating set M_o of $L/K(L^p)$ and subsets B_i in L, $i = 1,2,\ldots$, such that the conditions in (a) are satisfied, then $M_i = \bigcup_{j=i+1}^{\infty} B_j$ is such that $M_i^{p^i}$ is a relative p-base of $K(L^{p^i})/K$, $i = 0,1,\ldots$.

Proof. A. We first prove (b). Since $M_{i-1}^{p^{i-1}}$ is a relative p-base of $K(L^{p^{i-1}})/K$, we have for all $b \in B_i$ that $b^{p^{i-1}} \notin K(L^{p^i})(B_i^{p^{i-1}} - b^{p^{i-1}}, M_i^{p^{i-1}})$. Hence $b^{p^i} \notin K^p(L^{p^{i+1}})(B_i^{p^i} - b^{p^i}, M_i^{p^i})$. Thus $B_i^{p^i}$ is p-independent of $M_i^{p^i}$ in $K(L^{p^i})$. Also $M_i^{p^i}$ is p-independent in $K(L^{p^i})$. Therefor $M_{i-1}^{p^i} = \bigcup_{j=i}^{\infty} B_j^{p^i}$ is p-independent in $K(L^{p^i})$.

To prove (a), we first show that $\bigcup_{i=1}^{\infty} B_i = M_o$. For every $m \in M_o$, there exists a positive integer i such that $m^{p^i} \in K$. Hence $m \notin M_i$ since $M_i^{p^i} \cap K = \emptyset$. Thus $\bigcap_{i=0}^{\infty} M_i = \emptyset$ whence $M_o = \bigcup_{i=1}^{\infty} B_i$. Now, to prove (1), we observe that $B_i^{p^i} \subseteq K(L^{p^i}) = K(L^{p^{i+1}})(M_i^{p^i})$. To show (2), we first observe that (1) implies that for all $b \in B_i$, $b^{p^i} \in K(L^{p^{i+1}})(B_i - b, M_i)$ and (b) implies that $M_{i-1}^{p^i}$ is p-independent in $K(L^{p^i})$. Therefore, if we set $S = (M_i \cup (B_i - b)) \cup \{b^{p^i}\}$, we have that S is p-independent in $K(L^{p^i})(S) = K(L^{p^{i+1}})(B_i - b, M_i)$ by Proposition 1.7. Consequently, $b^{p^i} \notin K^p(L^{p^{i+2}})(B_i^p - b^p, M_i^p)$, whence $b^{p^{i-1}} \notin K(L^{p^{i+1}})(B_i - b, M_i)$.

To prove (c), consider any other chain say $N = N_0 \supseteq N_1 \supseteq \ldots \supseteq N_i \supseteq \ldots$, such that $N_i^{p^i}$ is a relative p-base of $K(L^{p^i})/K$, $i = 0,1,\ldots$. Set $C_i = N_{i-1} - N_i$, $i = 1,2,\ldots$. Fix i and set $K^* = K(L^{p^i})$. Then $K^* = K(K^{*p})(M_i^{p^i}) = K(K^{*p})(N_i^{p^i})$. Since $M_i^{p^i}$ is a relative p-base of K^*/K and $M_i^{p^i} \cup B_i^{p^i} = M_{i-1}^{p^i}$ is p-independent in K^* by (b), there exists a subset G_0 of K such that $G_0 \cup M_i^{p^i} \cup B_i^{p^i}$ is a p-base of K^*. Now $K^{*p}(M_i^{p^i}, B_i^{p^i})$ $= (K^*(M_{i-1}^{p^{i-1}}))^p = (K^*(L^{p^{i-1}}))^p = (K^*(N_{i-1}^{p^{i-1}}))^p = K^{*p}(N_i^{p^i}, C_i^{p^i})$. Therefore $G_0 \cup N_i^{p^i} \cup C_i^{p^i}$ is also a p-base of K^*. Since $M_i^{p^i}$ and $N_i^{p^i}$ are relative p-bases of K^*/K, there exists a subset G of K such that $G_0 \subseteq G$ and $G \cup M_i^{p^i}$ as well as $G \cup N_i^{p^i}$ are p-bases of K^*. We already know that $G_0 \cup M_i^{p^i} \cup B_i^{p^i}$ and $G_0 \cup N_i^{p^i} \cup C_i^{p^i}$ are p-bases of K^*. Hence $G - G_0$, $B_i^{p^i}$ and $G - G_0$, $C_i^{p^i}$ are each minimal generating sets of $K^*/K^{*p}(G_0, M_i^{p^i})$ and $K^*/K^{*p}(G_0, N_i^{p^i})$, respectively. Thus $|B_i| = |G - G_0| = |C_i|$. Now $|M_i| = |N_i|$ because $M_i^{p^i}$ and $N_i^{p^i}$ are relative p-bases of K^*/K.

B. For $j \le i$, $B_j^{p^j} \subseteq K(L^{p^{j+1}})(B_{j+1}^{p^j}, B_{j+2}^{p^j},\ldots)$ by (1). Thus

$i \subseteq K^{p^{i-j}}(L^{p^{i+1}})(B_{j+1}^{p^i}, B_{j+2}^{p^i},\ldots) \subseteq K(L^{p^{i+1}})(B_{j+1}^{p^i}, B_{j+2}^{p^i},\ldots)$.

nce $K(L^{p^i}) = K(L^{p^{i+1}})(M_i^{p^i})$. Therefore it remains to be shown

at $M_i^{p^i}$ is relatively p-independent in $K(L^{p^i})/K$. Suppose

i is not relatively p-independent. Then there exists $b \in M_i$

ch that $b^{p^i} \in K(L^{p^{i+1}})(M_i^{p^i} - b^{p^i})$. Let k be the smallest

sitive integer for which B_{i+k} contains such an element b.

$k = 1$, then (2) is contradicted. Suppose $k > 1$ and that

i cannot be represented as an element of $K(L^{p^{i+1}})(M_i^{p^i} - b^{p^i})$

thout using elements from $B_{i+1}^{p^i} \cup \ldots \cup B_{i+k-1}^{p^i}$. Then an element

om the latter set can be exchanged for b^{p^i}. However this

ntradicts the minimality of k. Thus $b^{p^i} \in K(L^{p^{i+1}})(B_{i+k}^{p^i} - b^{p^i}, M_{i+k}^{p^i})$

om which it follows that $b^{p^{i+k-1}} \in K(L^{p^{i+k}})(B_{i+k}^{p^{i+k-1}} - b^{p^{i+k-1}}, M_{i+k}^{p^{i+k-1}})$

$L^{p^{i+k+1}}, M_{i+k}^{p^{i+k}})(B_{i+k}^{p^{i+k-1}} - b^{p^{i+k-1}}, M_{i+k}^{p^{i+k-1}}) \subseteq K(L^{p^{i+k+1}})(B_{i+k} - b, M_{i+k})$.

this contradicts (2). Hence $M_i^{p^i}$ is a relative p-base of

$L^{p^i})/K$. q.e.d.

1.30. Definition. If L/K is purely inseparable and

$B_2,\ldots$ are subsets of L satisfying the conditions in (a) of

oposition 1.29, then $\{B_1, B_2,\ldots\}$ is called a canonical system

L/K and M_o is called canonically ordered.

1.31. Corollary. A. If L/K is purely inseparable and if there exist subsets M_i in L such that $M_i \supseteq M_{i+1}$ and $M_i^{p^i}$ is a minimal generating set of $K(L^{p^i})/K$, $i = 0,1,\ldots$, then the subsets defined by $B_i = M_{i-1} - M_i$, $i = 1,2,\ldots$, satisfy:

(a) $\bigcup_{i=1}^{\infty} B_i = M_o$ and for $i = 1,2,\ldots$,

(1) $B_i^{p^i} \subseteq K(B_{i+1}^{p^i}, B_{i+2}^{p^i},\ldots)$,

(2) For all $b \in B_i$, b has exponent i over $K(B_i - b, B_{i+1},\ldots)$.

Conditions (b) and (c) of Proposition 1.29 hold.

B. Given a generating set M_o of L/K and subsets B_i in L, $i = 1,2,\ldots$, such that the conditions in (a) are satisfied, then $M_i = \bigcup_{j=i+1}^{\infty} B_j$ is such that $M_i^{p^i}$ is a minimal generating set of $K(L^{p^i})/K$, $i = 0,1,\ldots$.

Proof. If $M_i^{p^i}$ is a minimal generating set of $K(L^{p^i})/K$, then $M_i^{p^i}$ is a relative p-base of $K(L^{p^i})/K$. For (1), (2) and B, note that if $M_i^{p^i}$ is a minimal generating set of $K(L^{p^i})/K$, then $K(B_{i+1}^{p^i}, B_{i+2}^{p^i},\ldots) = K(L^{p^{i+1}})(B_{i+1}^{p^i}, B_{i+2}^{p^i},\ldots)$ and $K(B_i - b, B_{i+1},\ldots) = K(L^{p^{i+1}})(B_i - b, B_{i+1},\ldots)$. q.e.d.

1.32. Definition. If L/K is purely inseparable and $B_1, B_2,\ldots$ are subsets of L satisfying the conditions in (a) of

Corollary 1.31, then $\{B_1, B_2, \ldots\}$ is called a canonical generating system of L/K, M_0 is said to be canonically ordered and L/K is said to be canonically generated. The elements of $\bigcup_{i=1}^{\infty} B_i$ are called canonical generators of L/K.

1.33. Example. The existence of a minimal generating set of L/K does not insure the existence of a canonical generating system of L/K: Let P be a perfect field and $z, x_1, x_2, \ldots$ independent indeterminates over P. Let $K = P(z, x_1, x_2, \ldots)$ and $L = K(x_1^{p^{-1}}, x_2^{p^{-1}}, \ldots)(z^{p^{-1}}, z^{p^{-2}}, \ldots)$. Then $\{x_i^{p^{-1}} z^{p^{-i}} \mid i = 1, 2, \ldots\}$ is a minimal generating set of L/K. However $K(L^p) = K(L^{p^2}) = \ldots$, whence $K(L^{p^i})/K$ can have no minimal generating set.

1.34. Proposition. Suppose L/K is purely inseparable and has a minimal generating set $M = M_0$. There always exist subsets M_i in M_0 such that $M_i \supseteq M_{i+1}$ and $M_i^{p^i}$ is a relative p-base of $K(L^{p^i})/K$. Then $M_i^{p^i}$ is a minimal generating set of $K(L^{p^i})/K$ if and only if $B_i = M_{i-1} - M_i$ is of bounded exponent over $K(M_i^{p^i})$, $i = 1, 2, \ldots$.

Proof. Suppose B_i is of bounded exponent over $K(M_i^{p^i})$ and $M_{i-1}^{p^{i-1}}$ is a minimal generating set of $K(L^{p^{i-1}})/K$ (the existence of M_0 being assumed). Then $K(L^{p^i}) = K(M_{i-1}^{p^i})$, and since $M_i^{p^i}$ is a relative p-base of $K(L^{p^i})/K$ we have $B_i^{p^i} \subseteq L^{p^i} \subseteq K(L^{p^{i+1}})(M_i^{p^i}) = K(M_{i-1}^{p^{i+1}})(M_i^{p^i}) = K(B_i^{p^{i+1}})(M_i^{p^i})$. Assume $B_i^{p^i} \subseteq K(B_i^{p^{i+n}})(M_i^{p^i}), n \geq 1$. Then $B_i^{p^{i+1}} \subseteq K(B_i^{p^{i+n+1}})(M_i^{p^i})$, whence $B_i^{p^i} \subseteq K(B_i^{p^{i+n+1}})(M_i^{p^i})$. Hence $B_i^{p^i} \subseteq K(M_i^{p^i})$. Thus $K(L^{p^i}) = K(M_i^{p^i})$. Therefore the desired result follows by induction. The converse is immediate. q.e.d.

1.35. Corollary. If L/K has an exponent, then every relative p-base of L/K yields a canonical generating system of L/K.

When a subset A of L is called a subbase over K, we mean that A is a subbase of $K(A)/K$. A subbase is called equi-exponential if every element in it has the same exponent.

If L/K has an exponent e, then Corollaries 1.31 and 1.35 and the following Proposition show that since $B_{e+1} = B_{e+2} = \dots = \emptyset$, $B_1, \dots, B_e$ can be chosen in the following manner: If $B_i \neq \emptyset$, then B_i is taken as a maximal subset of L with respect to the property that B_i is an equi-exponential subbase over $K(B_{i+1}, \dots, B_e)$ with exponent i and i is the exponent of $L/K(B_{i+1}, \dots, B_e)$, $i = 1, \dots, e$.

1.36. Proposition. Suppose L/K has an exponent e. The conditions in (a) of Corollary 1.31 are equivalent to the following conditions: If $B_i \neq \emptyset$, then for all $b \in B_i$, L has exponent i over $K(B_i - b, B_{i+1}, \dots, B_e)$ and i is the exponent of B_i over $K(B_i - b, B_{i+1}, \ldots, B_e)$.

Proof. Suppose the conditions in (a) of Corollary 1.31 hold. If $B_i \neq \emptyset$, then clearly for all $b \in B_i$, $L^{p^i} \subseteq K(B_i - b, B_{i+1}, \ldots, B_e)$ and L must have exponent i over $K(B_i - b, B_{i+1}, \ldots, B_e)$ since b has exponent i over $K(B_i - b, B_{i+1}, \ldots, B_e)$. Conversely suppose the conditions of the Proposition hold. Then (2) is immediate. Now $L^{p^i} \subseteq \bigcap_{b \in B_i} K(B_i - b, B_{i+1}, \ldots, B_e) = K(B_{i+1}, \ldots, B_e)$. Thus $B_i^{p^i} \subseteq K(M_i)(B_{i+1}^{p^i})$. Let $K' = K(M_j)(B_{i+1}^{p^i}, \ldots, B_j^{p^i})$ and $q = p^{j-i}$. Assume $B_i^{p^i} \subseteq K'(B_j)$. If we show that this implies $B_i^{p^i} \subseteq K'$, then (1) holds by induction. Now for $b \in B_i$, $b^{p^i} = \sum k_{t_1 \ldots t_s} b_1^{t_1} \ldots b_s^{t_s}$, where $k_{t_1 \ldots t_s} \in K'$, $b_1, \ldots, b_s \in B_j$ and $0 \leq t_1, \ldots, t_s < p^i$. Thus $b^{p^j} = (b^{p^i})^q = \sum k^q_{t_1 \ldots t_s} (b_1^q)^{t_1} \ldots (b_s^q)^{t_s}$. Since L has exponent $\leq j$ over $K(M_j)$, $B_i^{p^j} \subseteq K(M_j)$, $i = 1, \ldots, e$. Thus $k^q_{t_1 \ldots t_s} \in K^q(M_j^q)(B_1^{p^j}, \ldots, B_j^{p^j})$ implies $k^q_{t_1 \ldots t_s} \in K(M_j)$. By (2), the products $(b_1^q)^{t_1} \ldots (b_s^q)^{t_s}$ form a linear basis of $K(M_j)(b_1^q, \ldots, b_s^q)/K(M_j)$. Since b^{p^j} and $k^q_{t_1 \ldots t_s} \in K(M_j)$, $k^q_{t_1 \ldots t_s} = 0$ for $t_1 + \ldots + t_s > 0$. Hence $b^{p^i} = (b^{p^j})^{q^{-1}} \in K'$. q.e.d.

If empty B_i's of Proposition 1.29 or Corollary 1.31 are deleted and the remaining ones are relabeled keeping them in the same order, then we get similar results by making the substitutions p^{e_i} for p^i and B_i for B_{e_i}, where B_{e_i} is non-empty. In this case we have $e_1 < e_2 < \ldots$, and e_i and the cardinalities of B_i, $\{B_i\}$ and M_{i-1}, $i = 1,2,\ldots$, are invariants of the extension.

<u>1.37. Proposition</u>. Let L/K be purely inseparable and let M_o be a generating set of L/K. Suppose there exist non-empty disjoint subsets B_i of M_o such that for distinct positive integers e_i, $i = 1,2,\ldots$:

(a) $\bigcup_{i=1}^{\infty} B_i = M_o$ and for $i = 1,2,\ldots$,

(1) $B_i^{q_i} \subseteq K(B_{i+1}^{q_i}, B_{i+2}^{q_i},\ldots)$ where $q_i = p^{e_i}$,

(2) For all $b \in B_i$, b has exponent e_i over $K(B_i - b, B_{i+1},\ldots)$.

Then there exists a reordering of the B_i with the same e_i as in (1) and (2) such that (1), (2) and also (3) $e_1 < e_2 < \ldots$ hold with the new ordering.

Proof. Reorder $B_1, B_2,\ldots$ as $B_{j_1}, B_{j_2},\ldots$ so that with the same e_i's, $e_{j_1} < e_{j_2} < \ldots$. Let $M_i' = B_{j_{i+1}} \cup B_{j_{i+2}} \cup \ldots$ and $M_i = B_{i+1} \cup B_{i+2} \cup \ldots$. Since $B_1 \cup B_2 \cup \ldots$ satisfies (1), we have for any $i = 1,2,\ldots$ that $B_{j_i}^{q_{j_i}} \subseteq K(M_{j_i}^{q_{j_i}}) \subseteq$

$K(M_i'^{q_{j_i}})(M_{i_j}^{q_{j_i}} - M_i'^{q_{j_i}})$. Now for $t = j_i + 1, j_i + 2, \ldots,$ $q_t \leq q_{j_i}$ when $B_t \nsubseteq M_i'$. Now, $B_t^{q_t} \subseteq K(M_t^{q_t})$ since $B_1 \cup B_2 \cup \ldots$ satisfies (1). Thus, when $B_t \nsubseteq M_i'$, $B_t^{q_{j_i}} \subseteq K^{q_{j_i}/q_t}(M_t^{q_{j_i}}) \subseteq K(M_t^{q_{j_i}})$. Hence $K(M_i'^{q_{j_i}})(M_{j_i}^{q_{j_i}} - M_i'^{q_{j_i}})$ $= K(M_i'^{q_{j_i}})$ since there are only a finite number of $B_t \nsubseteq M_i'$. Thus $B_{j_i}^{q_{j_i}} \subseteq K(M_i'^{q_{j_i}})$, so $\bigcup_{i=1}^{\infty} B_{j_i}$ satisfies (1), (3). Since $\bigcup_{i=1}^{\infty} B_i$ satisfies (2), we have for every finite subset $B_i' \subseteq B_i$ with m_i elements and for any $r = 1,2,\ldots$ the degree relation $[K_r(B_1',\ldots,B_r') : K_r] \geq p^{e_1 m_1} \ldots p^{e_r m_r}$, where $K_r = K(M_r)$ and $i = 1,2,\ldots$. Suppose there exists $b \in B_{j_i}$ such that $b^{q_{j_i}/p} \in K(M_i')(B_{j_i} - b)$. Let $r = j_i$. Then $B_{j_i} \cap K = \emptyset$ and there exists $s \geq 1$ such that $B_{j_{i+s}} \cup B_{j_{i+s+1}} \cup \ldots \subseteq K_r$ since K_r contains all but a finite number of the B_i's. Then $b^{q_{j_i}/p} \in K_r(B_{j_{i+1}},\ldots,B_{j_{i+s-1}})(B_{j_i} - b)$ since $K_r(B_{j_{i+1}},\ldots,B_{j_{i+s-1}}) \supseteq K(M_i')$. Thus there exist finite sets $B_{j_t}'' \subseteq B_{j_t}$ $(t = i+1,\ldots,i+s-1)$ and $B_{j_i}'' \subseteq B_{j_i} - \{b\}$ such that $b^{q_{j_i}/p} \in K_r(B_{j_{i+1}}'',\ldots,B_{j_{i+s-1}}'')(B_{j_i}'')$, where $B_{j_t}'' = \emptyset$ if

$B_{j_t} \subseteq K_r$. Since $B_{j_1} \cup B_{j_2} \cup \ldots$ satisfies (1) and $K_r \supseteq K(M'_{j_{i+s-1}})$, there exist finite sets $B'_{j_t} \supseteq B''_{j_t}$ $(t = i + 1,\ldots,i + s - 1)$ such that $b^{q_{j_i}/p} \in K_r(B'_{j_{i+1}},\ldots,B'_{j_{i+s-1}})(B''_{j_i})$, $B''^{\,q_{j_i}}_{j_i} \subseteq K_r(B'_{j_{i+1}},\ldots,B'_{j_{i+s-1}})$ and $B'^{\,q_{j_t}}_{j_t} \subseteq K_r(B'_{j_{t+1}},\ldots,B'_{j_{i+s-1}})$ $(t = 1 + 1,\ldots,i + s - 1)$, where $B'_{j_t} = \emptyset$ if $B_{j_t} \subseteq K_r$. However this contradicts the above degree relation. q.e.d.

In the above we deduced invariants for L/K by use of the intermediate fields $K(L^{p^i})$, $i = 1,2,\ldots$. By considering the intermediate fields $(K^{p^{-j}} \cap L)(L^{p^i})$, we shall derive additional invariants associated with a subclass of canonical generating systems. But first we derive invariants for L/K by use of the intermediate fields $K^{p^{-j}} \cap L$, $j = 1,2,\ldots$.

Set $L_j = K^{p^{-j}} \cap L$, $j = 0,1,\ldots$ and for the composite of L_{i-1} with $L_j^{p^{j-1}}$, set $L_{ij} = L_{i-1}L_j^{p^{j-i}}$, $j \geq i = 1,2,\ldots$. For each positive integer j, let T_{jj} be a relative p-base of $L_j/L_{j,j+1}$. Making use of $L_{j-1}^p \subseteq L_{j-2}$, we find that if $j > 1$, then $L_{j-1,j} = L_{j-2}L_j^p = L_{j-2}L_{j,j+1}^p(T_{jj}^p) = L_{j-1,j+1}(T_{jj}^p)$. Hence we can select from T_{jj}^p a relative p-base $T_{j-1,j}$ of $L_{j-1,j}/L_{j-1,j+1}$. Now, with j fixed, we continue to select relative p-bases as follows: Let $k < j$ and suppose we have a relative p-base T_{kj} of $L_{kj}/L_{k,j+1}$ that is contained in $T^p_{k+1,j}$. Then using the fact that $L^p_{k-1} \subseteq L_{k-2}$, we have

$$L_{k-1,j} = L_{k-2}L_j^{p^{j-k+1}} = L_{k-2}L^p_{kj} = L_{k-2}L^p_{k,j+1}(T^p_{kj}) = L_{k-1,j+1}(T^p_{kj}).$$

Hence we can select from T^p_{kj} a relative p-base $T_{k-1,j}$ of $L_{k-1,j}/L_{k-1,j+1}$. For each j, this process ends after a finite number of steps, namely, with the construction of T_{1j}. We can think of this process as filling boxes in an upper triangular infinite grid. When L/K has an exponent e, then $T_{kj} = \emptyset$ for $j \geq k > e$.

Now define $T_k = \bigcup_{j=k}^{\infty} T_{kj}$. Since T_{kj} is a relative p-base of $L_{kj}/L_{k,j+1}$ and $L_k \supseteq L_{k,k+1} \supseteq \cdots \supseteq L_{kj} \supseteq L_{k,j+1} \supseteq \cdots \supseteq L_{k-1}$, it follows that T_k is relatively p-independent in L_k/L_{k-1}. For convenience, we set $N = \bigcup_{j=1}^{\infty} T_{jj}$ and $N_{ij} = T_{ij}^{p^{i-j}}$ $(j \geq i = 1,2,\ldots)$, so $T_{jj} = N_{jj} \supseteq N_{j-1,j} \supseteq \cdots \supseteq N_{1j}$, $j = 1,2,\ldots$, and $N_{ii} \cap N_{jj} = \emptyset$ when $i \neq j$.

1.38. <u>Definition</u>. The collection of subsets T_{ij} constructed for L/K in the above fashion is called a lower tower system of L/K and is denoted by $\{T_{ij}\}$. The subset $N = \bigcup_{j=1}^{\infty} T_{jj}$ is called a lower tower set of L/K.

From the definition of T_{ij} and N it follows that $|T_{ij}|$ and $|N|$ are invariants of L/K.

1.39. Proposition. Let $\{T_{ij}\}$ be a lower tower system of L/K. Then $|N_{i+j,j} - N_{ij}|$ is an invariant of L/K, $1 \leq i < j = 1,2,\ldots$.

Proof. Consider $L_{ij} = L_{i,j+1}(T_{ij})$ as a composite of $L^p_{i+1,j}$ and $L_{i,j+1}$ over $L^p_{i+1,j+1}$. Let A be a relative p-base of $L_{i,j+1}/L^p_{i+1,j+1}$. Then there is a set $A^* \subseteq A$ such that A^* is a relative p-base of $L_{ij}/L^p_{i+1,j}$. Since $L^p_{ij} \subseteq L^p_{i+1,j+1}$, both $(A^* \cup T_{ij}) \cup (A - A^*)$ and $(A^* \cup T_{ij}) \cup (T^p_{i+1,j} - T_{ij})$ are relative p-bases of $L_{ij}/L^p_{i+1,j+1}$. Since $A - A^*$ is independent of the lower tower system and the cardinality of a relative p-base is unique, we deduce that $|N_{i+1,j} - N_{ij}| = |T^p_{i+1,j} - T_{ij}| = |A - A^*|$ for every choice of $\{T_{ij}\}$. q.e.d.

1.40. Definition. If $x \in T_{jj}$, then its length $\ell(x)$ is defined by $\ell(x) = \max_k\{k \mid x^{p^k} \in T_{j-k,j}\}$. The length of T_{jj}, denoted by ℓ_j, is defined by $\ell_j = \max\{\ell(x) \mid x \in T_{jj}\}$.

By the definition of ℓ_j, $N_{j-\ell_j,j}$ is the last non-empty term in the chain $N_{jj} \supseteq \ldots \supseteq N_{j-\ell_j,j} \supseteq \cdots \supseteq N_{1j}$. Also, the invariant $|N_{i+1,j} - N_{ij}|$ is the number of elements in T_{jj} with length $j - i - 1$.

1.41. Definition. Let $\{T_{ij}\}$ be a lower tower system of L/K. The numbers $|N|, |N_{ij}|, |N_{i+1,j} - N_{ij}|$, ℓ_j are called the lower tower invariants of L/K.

<u>1.42. Proposition</u>. If $\{T_{ij}\}$ and $\{T'_{ij}\}$ are any two lower tower systems of L/K, then there exists a $1-1$ mapping g of N onto N' such that

(1) $g(N_{ij}) = N'_{ij}$, $1 \leq i \leq j = 1,2,\ldots$ and

(2) $\ell(x) = \ell(g(x))$ for every $x \in N$.

Proof. Fix j and note that $|N_{ij}| = |N'_{ij}|$ for $i \leq j$ and $|N_{i+1,j} - N_{ij}| = |N'_{i+1,j} - N'_{ij}|$ for $i = j - \ell_j,\ldots,j-1$. Hence when $i < j$ any $1-1$ mapping of N_{ij} onto N'_{ij} can be extended to a $1-1$ mapping of $N_{i+1,j}$ onto $N'_{i+1,j}$. Starting with $N_{j-\ell_j,j}$, we can construct a $1-1$ mapping g_j from N_{jj} onto N'_{jj} such that $g_j | N_{ij}$ maps N_{ij} $1-1$ onto N'_{ij}, $i = j - \ell_j,\ldots,j$. This mapping automatically preserves lengths. The required mapping g is now defined to be the union of the g_j's. q.e.d.

<u>1.43. Definition</u>. A lower tower system $\{T_{ij}\}$ is called a lower tower generating system of L/K if and only if $L_j = L_{j-1}(T_j)$, $j = 1,2,\ldots$, and N is called a lower tower generating set of L/K if and only if $L = K(N)$.

<u>1.44. Proposition</u>. (a) If L/K has bounded exponent, then every lower tower system (set) is a lower tower generating system (set).

(b) If L/K has bounded exponent, then the set of monomials of the following type: $\prod_{x \in N} x^{e_x}$, $0 \leq e_x < p^{\ell(x)+1}$, where all but a finite number of the e_x are zero, is a linear basis of L/K.

Proof. (a) In this case, T_j is a relative p-base of L_j/L_{j-1} for every j.

(b) By part (a), the monomials $\prod_{x \in T_j} x^{e_x}$, $0 \leq e_x < p$, where all but a finite number of the e_x are zero, form a linear basis of L_j/L_{j-1}. Since we have the sequence of fields $L_e \supset \ldots \supset L_o$, L/K has a linear basis consisting of the basis elements of the intermediate extensions. Thus, combining exponents, L/K has the given linear basis. q.e.d.

For $x \in T_{jj}$, the height of x is defined to be the integer $h(x) = j - 1$.

Let N denote a lower tower set $\bigcup_{j=1}^{\infty} T_{jj}$ of L/K. Since T_{jj} is a relative p-base of $L_j / L_{j-1}(L_{j+1}^p)$, T_{jj} contains a relative p-base of $L_j(L^p) / L_{j-1}(L^p)$, $j = 1,2,\ldots$. Thus N contains a relative p-base of L/K.

1.45. Example. A lower tower generating set of L/K need not be a minimal generating set of L/K: Let $L = K(z^{p^{-3}}, z^{p^{-3}}x^{p^{-1}} + y^{p^{-2}})$, where $K = P(x,y,z)$, P is a perfect field and x,y,z are independent indeterminates over P. Then $\{T_{ij}\}$ is a lower tower system of L/K where: $T_{33} = \{z^{p^{-3}}, z^{p^{-3}}x^{p^{-1}} + y^{p^{-2}}\}$, $T_{23} = \{z^{p^{-2}}\}$, $T_{13} = \{z^{p^{-1}}\}$, $T_{22} = T_{12} = \emptyset$, $T_{11} = \{y^{p^{-1}}\}$. Hence $\{z^{p^{-3}}, z^{p^{-3}}x^{p^{-1}} + y^{p^{-2}}, y^{p^{-1}}\}$

s a lower tower generating set of L/K while $\{z^{p^{-3}}$,
$^{p^{-3}}x^{p^{-1}} + y^{p^{-2}}\}$ is a minimal generating set of L/K.

1.46. Proposition. The following conditions are equivalent.

(1) Every lower tower set N of L/K is a relative -base of L/K.

(2) There exists a lower tower set N of L/K which s a relative p-base of L/K.

(3) L_j and $L_{j-1}(L^p)$ are linearly disjoint over $_{j-1}(L^p_{j+1})$, $j = 1,2,\dots$.

Proof. (1) implies (2) immediately. Suppose (2) holds. ince T_{jj} is a relative p-base of $L_j/L_{j-1}(L^p_{j+1})$, T_{jj} ontains a relative p-base of $L_j(L^p)/L_{j-1}(L^p)$, $j = 1,2,\dots$. y (2), T_{11} must be a relative p-base of $L_1(L^p)/K(L^p)$. ince $L_1/K(L^p_2)$ has exponent 1, (3) holds if $j = 1$. Let it hold for $= 1,\dots,i$. Then T_{jj} is a relative p-base of $L_j(L^p)/L_{j-1}(L^p)$ or $j = 1,\dots,i$. If $T_{i+1,i+1}$ is not a relative p-base of $_{i+1}(L^p)/L_i(L^p)$, then we have that $\bigcup_{j=1}^{i+1} T_{jj}$ is a relatively -dependent set in L/K which contradicts (2). Thus $T_{i+1,i+1}$ s a relative p-base of $L_{i+1}(L^p)/L_i(L^p)$. Since $L_{i+1}/L_i(L^p_{i+2})$ as exponent 1, (3) holds for $j = i + 1$. Hence (3) holds by nduction. Suppose (3) holds and let N be any tower set of $/K$. Then T_{jj} is a relative p-base of $L_j(L^p)/L_{j-1}(L^p)$, $= 1,2,\dots$. Thus N is relatively p-independent in L/K.

Clearly $L = K(L^p)(N)$. Therefore (1) holds. q.e.d.

We have refined the chain of fields $K \subseteq K^{p^{-1}} \cap L \subseteq \dots \subseteq K^{p^{-i}} \cap L \subseteq \dots$, $i = 0,1,\dots$, by introducing the fields

(1) $(K^{p^{-i+1}} \cap L)(K^{p^{-i}} \cap L^{p^j})$, $j = 1,2,\dots$, between $K^{p^{-i+1}} \cap L$ and $K^{p^{-i}} \cap L$. We now refine the chain of fields $L \supseteq K(L^p) \supseteq \dots \supseteq K(L^{p^i}) \supseteq \dots$, $i = 0,1,\dots$, by introducing the fields

(2) $K(L^{p^i})(K^{p^{-j}} \cap L^{p^{i-1}})$, $j = 1,2,\dots$, between $K(L^{p^i})$ and $K(L^{p^{i-1}})$.

1.47. Definition. The ascending chain of fields in (1) is called the lower tower of L/K and the descending chain of fields in (2) is called the upper tower of L/K.

We now give a construction of the relative p-bases of the upper tower of fields. Let M_{1j} be a relative p-base of $K(L^p)(L_j)/K(L^p)(L_{j-1})$, $j = 1,2,\dots$. Then $M_o = \bigcup_{j=1}^{\infty} M_{1j}$ is a relative p-base of $L/K(L^p)$ and L/K. Now M_{1j}^p clearly contains a relative p-base of $K(L^{p^2})(L_j^p)/K(L^{p^2})(L_{j-1}^p)$, say M_{2j}, $j = 1,2,\dots$, $M_{21} = \emptyset$. Then $M_1^p = \bigcup_{j=2}^{\infty} M_{2j}$ is a relative p-base of $K(L^p)/K(L^{p^2})$ and $K(L^p)/K$. Suppose $M_{i-1}^{p^{i-1}} = \bigcup_{j=i}^{\infty} M_{ij}$ is a relative p-base of $K(L^{p^{i-1}})/K(L^{p^i})$ where M_{ij} is a relative p-base of $K(L^{p^i})(L_j^{p^{i-1}})/K(L^{p^i})(L_{j-1}^{p^{i-1}})$. Now M_{ij}^p contains a relative p-base of $K(L^{p^{i+1}})(L_j^{p^i})/K(L^{p^{i+1}})(L_{j-1}^{p^i})$, say $M_{i+1,j}$, $j = i, i+1, \ldots, M_{i+1,i} = \emptyset$. Then $M_i^{p^i} = \bigcup_{j=i+1}^{\infty} M_{i+1,j}$ is a relative p-base of $K(L^{p^i})/K(L^{p^{i+1}})$ and $K(L^{p^i})/K$. The construction of the M_{ij}'s can be thought of as filling boxes in an upper triangular grid.

When L/K is purely inseparable we have that $\{B_1, B_2, \ldots\}$ is a canonical system of L/K where $B_i = M_{i-1} - M_i$ $= \bigcup_{j=i}^{\infty} (M_{ij}^{p^{-i+1}} - M_{i+1,j}^{p^{-i}})$, $i = 1,2,\ldots$. When L/K has exponent e, $M_{i-1}^{p^{i-1}} = M_{ii} \cup \ldots \cup M_{ie}$, $i = 1,\ldots,e$.

<u>1.48. Definition</u>. The collection of subsets $\{M_{ij}\}$ constructed in the above fashion for the upper tower of fields is called an upper tower system of L/K.

<u>1.49. Definition</u>. For $x \in M_{1j}$ the depth of x denoted by $d(x)$ is defined by $d(x) = \max_k \{x^{p^k} \mid x^{p^k} \in M_{1+k,j}\}$, $j = 1,2,\ldots$. (Since $M_{j+1,j} = \emptyset$, we have $d(x) \leq j - 1$.) The depth of M_{1j} denoted by d_j is defined by $d_j = \max\{d(x) \mid x \in M_{1j}\}$.

<u>1.50 Proposition</u>. (a) For an upper tower system $\{M_{ij}\}$ of L/K, $|M_{ij}|$, $|M_{ij}^p - M_{i+1,j}|$ and d_j are invariants of the extension. (We call these the upper tower invariants of L/K.)

(b) If $\{M_{ij}\}$ and $\{M'_{ij}\}$ are any two upper tower systems of L/K, then there exists a $1 - 1$ mapping g of M_o onto

M'_o such that $g(M_{ij}) = M'_{ij}$ and $d(x) = d(g(x))$ for every $x \in M_o$.

Proof. For the proof of (a), let A be a relative p-base of $K(L^{p^{i+1}})(L_{j-1}^{p^i})/K^p(L^{p^{i+1}})(L_{j-1}^{p^i})$ and let A^* be a subset of A such that A^* is a relative p-base of $K(L^{p^{i+1}})(L_j^{p^i})/K^p(L^{p^{i+1}})(L_j^{p^i})$. Using the fact that $K(L^{p^{i+1}})(L_j^{p^i})/K^p(L^{p^{i+1}})(L_{j-1}^{p^i})$ has exponent 1, we have that $M_{i+1,j} \cup A = (M_{i+1,j} \cup A^*) \cup (A - A^*)$ and $M_{ij}^p \cup A^* = (M_{i+1,j} \cup A^*) \cup (M_{ij}^p - M_{i+1,j})$ are relative p-bases of $K(L^{p^{i+1}})(L_j^{p^i})/K^p(L^{p^{i+1}})(L_{j-1}^{p^i})$. Hence $|A - A^*| = |M_{ij}^p - M_{i+1,j}|$. Since $A - A^*$ is independent of the upper tower system, we have the invariance of $|M_{ij}^p - M_{i+1,j}|$ and thus the invariance of d_j. The invariance of $|M_{ij}|$ is immediate. Arguments similar to those used in Proposition 1.42 show that (b) holds. q.e.d.

In the following we show that when L/K has an exponent, there is a lower tower system which can be used in some order for an upper tower system if and only if certain invariantly described pairs of intermediate fields are linearly disjoint. This gives a relation between canonical generating systems and lower tower generating systems.

1.51. Proposition. Suppose L/K has exponent e. Then the following conditions are equivalent for $i = 1,\ldots,e$.

(1) For all lower tower systems $\{T_{ij}\}$ of L/K, $K(L^{p^{e-i}}) = K(\bigcup_{j=1}^{i} T_{j,e-i+j})$.

(2) There is a lower tower system $\{T_{ij}\}$ of L/K such that $K(L^{p^{e-i}}) = K(\bigcup_{j=1}^{i} T_{j,e-i+j})$.

(3) $K(L_{e-j}^{p^{e-i}})$ and $L_{i-j-1}(L_{e-j+1}^{p^{e-i+1}})$ are linearly disjoint over $K(L_{e-j-1}^{p^{e-i}})(L_{e-j+1}^{p^{e-i+1}})$, $j = 0,\ldots,i-1$.

Proof. Fix i. Clearly (1) implies (2). Suppose (2) holds. In conjunction with showing (3) holds, we also show that the following condition holds:

$$(*)\quad K(L_{e-j}^{p^{e-i}}) = K(T_{i-j,e-j},\ldots,T_{i-j-s,e-j-s},\ldots,T_{1,e-i+1})$$
$$(T^{p}_{i-j+1,e-j+1},\ldots,T^{p^{t+1}}_{i-j+1+t,e-j+1+t},\ldots,T^{p^{j}}_{ie}),$$

$j = 0,\ldots,i-1;\ s = 0,\ldots,i-j-1;\ t = 0,\ldots,j-1.$

For $j = 0$, (*) is (2) of the proposition. Consider (3) for $j = 0$. Then $K(L_{e-1}^{p^{e-i}})(L_{e+1}^{p^{e-i+1}}) = K(L_{e-1}^{p^{e-i}})$ since $L_{e-1}^{p^{e-i}} \supseteq L^{p^{e-i+1}} \supseteq L_{e+1}^{p^{e-i+1}}$. Now T_{ie} is a relative p-base of $L_{i-1}(L^{p^{e-i}})/L_{i-1}(L_{e+1}^{p^{e-i+1}})$. Hence T_{ie} is relatively p-independent in $K(L^{p^{e-i}})/K(L_{e-1}^{p^{e-i}})$. Since $K(L^{p^{e-i}}) = K(\bigcup_{k=1}^{i} T_{k,e-i+k})$ and $K(L_{e-1}^{p^{e-i}}) \supseteq \bigcup_{k=1}^{i-1} T_{k,e-i+k}$, we have that T_{ie} is a relative p-base of $K(L^{p^{e-i}})/K(L_{e-1}^{p^{e-i}})$. Since this latter extension has exponent 1, (3) holds for $j = 0$. For n an integer such that $1 \leq n \leq i-1$ make the induction hypothesis that for

$j = 0,\ldots,n-1$ conditions (3) and (*) hold. Consider $j = n$. By the induction hypothesis, $K(L_{e-n+1}^{p^{e-i}}) = K(S)(S')$ where $S = T_{1,e-i+1} \cup \cdots \cup T_{i-n,e-n}$ and $T_{i-n+1,e-n+1} \subseteq S' \subseteq T_{i-n+1,e-n+1} \cup \cdots \cup T_{ie}^{p^{n-1}}$ so that S' is a relative p-base of $K(L_{e-n+1}^{p^{e-i}})/K(L_{e-n}^{p^{e-i}})$. (We note that S' exists since $T_{i-n+1,e-n+1} \cup \cdots \cup T_{ie}^{p^{n-1}}$ must generate $K(L_{e-n+1}^{p^{e-i}})/K(L_{e-n}^{p^{e-i}})$.) Now $K(L_{e-n+1}^{p^{e-i}}) \supseteq K(L_{e-n}^{p^{e-i}}) \supseteq K(S,S'^{p})$ and $K(L_{e-n+1}^{p^{e-i}})/K(S,S'^{p})$ has exponent 1 since this extension is generated by S'. Hence since S' is a relative p-base of $K(L_{e-n+1}^{p^{e-i}})/K(L_{e-n}^{p^{e-i}})$ and thus of $K(L_{e-n+1}^{p^{e-i}})/K(S,S'^{p})$, we have that $K(L_{e-n}^{p^{e-i}}) = K(S,S'^{p})$. This proves (*) for $j = n$. Now $K(L_{e-n-1}^{p^{e-i}})(L_{e-n+1}^{p^{e-i+1}}) \supseteq (S \cup S'^{p}) - T_{i-n,e-n}$. Hence $T_{i-n,e-n}$ must generate $K(L_{e-n}^{p^{e-i}})/K(L_{e-n-1}^{p^{e-i}})(L_{e-n+1}^{p^{e-i+1}})$. Since $T_{i-n,e-n}$ is a relative p-base of $L_{i-n-1}(L_{e-n}^{p^{e-i}})/L_{i-n-1}(L_{e-n+1}^{p^{e-i+1}})$ by definition of lower tower system, $T_{i-n,e-n}$ is certainly relatively p-independent in $K(L_{e-n}^{p^{e-n}})/K(L_{e-n-1}^{p^{e-i}})(L_{e-n+1}^{p^{e-i+1}})$. Thus $T_{i-n,e-n}$ is a relative p-base of $K(L_{e-n}^{p^{e-i}})/K(L_{e-n-1}^{p^{e-i}})(L_{e-n+1}^{p^{e-i+1}})$. Since this latter extension has exponent 1, (3) holds for $j = n$. Hence by induction, (3) and (*) hold for $j = 0,\ldots,i-1$.

Suppose (3) holds and let $\{T_{ij}\}$ be a lower tower system L/K. Then it follows that $T_{i-j,e-j}$ is a relative p-base of $K(L_{e-j}^{p^{e-i}})/K(L_{e-j-1}^{p^{e-i}})(L_{e-j+1}^{p^{e-i+1}})$ for $j = 0,\ldots,i-1$. It also follows that $T^{p}_{i-j+1,e-j+1} \cup \ldots \cup T^{p^{j}}_{ie}$ generates $K(L_{e-j-1}^{p^{e-i}})(L_{e-j+1}^{p^{e-i+1}})/K(L_{e-j-1}^{p^{e-i}})$, $j = 0,\ldots,i-1$. Thus $T_{1,e-i+1} \cup \ldots \cup T_{ie}$ generates $K(L^{p^{e-i}})/K(L_{e-i+2}^{p^{e-i+1}})$. Now $K(L_{e-i+2}^{p^{e-i+1}}) = K(\bigcup_{j=1}^{i} T^{p^{j-1}}_{j,e-i+j})$ by definition of lower tower system. Hence $K(L^{p^{e-i}}) = K(\bigcup_{j=1}^{i} T_{j,e-i+j})$. That is, (1) holds. q.e.d.

1.52. Proposition. Suppose L/K has exponent e. Let $\{T_{ij}\}$ be a lower tower system of L/K. The following conditions are equivalent.

(1) $B_{e-i+1} = \bigcup_{j=1}^{i} (T^{p^{-e+1}}_{j,e-i+j} - T^{p^{-e+i-1}}_{j-1,e-i+j})$, $T_{0i} = \emptyset$,

$i = 1,\ldots,e$, is such that $\{B_1,\ldots,B_e\}$ is a canonical generating system of L/K. (That is, B_{e-i+1} can be taken as $\{x \mid x \in N,\ \ell(x) = e - i\}$ where $\ell(x)$ denotes the length of x and $N = \bigcup_{j=1}^{e} T_{jj}$.).

(2) $T_{i-j,e-j}$ is a relative p-base of

$K(L^{p^{e-i+1}})(L_{e-j}^{p^{e-i}})/K(L^{p^{e-i+1}})(L_{e-j-1}^{p^{e-i}})$, $i = 1,\ldots,e$; $j = 0,\ldots,i-1$. (That is, $T_{i-j,e-j}$ can be taken as an $M_{e-i+1,e-j}$.)

(3) $K(L^{p^{e-i}}) = K(\bigcup_{j=1}^{i} T_{j,e-i+j})$, $i = 1,\ldots,e$

Proof. (1) implies (2): By Corollary 1.31, $\bigcup_{j=0}^{i-1} T_{i-j,e-j}$ $= \bigcup_{j=1}^{i} T_{j,e-i+j} = B_{e-i+1}^{p^{e-i}} \cup \ldots \cup B_{e}^{p^{e-i}}$ is a relative p-base of $K(L^{p^{e-i}})/K$. Since $K(L^{p^{e-i}})/K$ has an exponent, $K(L^{p^{e-i}})$ $= K(\bigcup_{j=1}^{i} T_{j,e-i+j})$. Hence (3) of Proposition 1.51. Since $T_{i-j,e-j}$ is thus a relative p-base of $K(L_{e-j}^{p^{e-i}})/K(L_{e-j-1}^{p^{e-i}})(L_{e-j+1}^{p^{e-i+1}})$, $T_{i-j,e-j}$ clearly contains a relative p-base of $K(L^{p^{e-i+1}})(L_{e-j}^{p^{e-i}})/K(L^{p^{e-i+1}})(L_{e-j-1}^{p^{e-i}})$. Since $\bigcup_{j=1}^{i} T_{j,e-i+j}$ is a relative p-base of $K(L^{p^{e-i}})/K$ and thus of $K(L^{p^{e-i}})/K(L^{p^{e-i+1}})$, $T_{i-j,e-j}$ is relatively p-independent in $K(L^{p^{e-i+1}})(L_{e-j}^{p^{e-i}})/K(L^{p^{e-i+1}})(L_{e-j-1}^{p^{e-i}})$.

(2) implies (3): It follows that $\bigcup_{j=1}^{i} T_{j,e-i+j}$ is a relative p-base of $K(L^{p^{e-i}})/K(L^{p^{e-i+1}})$ and thus of $K(L^{p^{e-i}})/K$. Hence $K(L^{p^{e-i}}) = K(\bigcup_{j=1}^{i} T_{j,e-i+j})$.

(3) implies (1): Now $B_{e-i+1}^{p^{e-i+1}} \subseteq K(L^{p^{e-i+1}}) \subseteq K(B_{e-i+2},\ldots,B_e)$ $i = 1,\ldots,e$. Thus for all $x \in B_{e-i+1}$, x has exponent

$\leq e - i + 1$ over $K(B_{e-i+1} - x, B_{e-i+2}, \ldots, B_e)$. By Proposition 1.44, every element x in B_{e-i+1} has exponent exactly $e - i + 1$ over $K(B_{e-i+1} - x, B_{e-i+2}, \ldots, B_e)$. Hence $\{B_1, \ldots, B_e\}$ satisfies the conditions in (a) of Corollary 1.31. By Corollary 1.31 B, $\{B_1, \ldots, B_e\}$ is a canonical generating system of L/K

q.e.d.

By Proposition 1.51 there exists a lower tower system satisfying any of the conditions (1), (2), (3) of Proposition 1.52 if and only if every lower tower system satisfies any of the conditions (1), (2), (3) of Proposition 1.52. Also for a lower tower system $\{T_{ij}\}$ of L/K, $K(L^{p^{e-i}}) = K(\bigcup_{j=1}^{i} T_{j,e-i+j})$, $i = 1, \ldots, e$, if and only if $\bigcup_{j=1}^{i} T_{j,e-i+j}$ is a relative p-base of $K(L^{p^{e-i}})/K$, $i = 1, \ldots, e$.

<u>1.53. Corollary</u>. Suppose L/K has exponent e. Then (3) of Proposition 1.51 with $i = 1, \ldots, e$ implies the following:

$K(L^{p^{e-i+1}})(L_{e-j-1}^{p^{e-i}})$ and $K(L_{e-j}^{p^{e-i}})$ are linearly disjoint over $K(L_{e-j-1}^{p^{e-i}})(L_{e-j+1}^{p^{e-i+1}})$, $i = 1, \ldots, e$; $j = 0, \ldots, i - 1$.

Proof. This implication follows from (2) of Proposition 1.52 and the fact that $K(L_{e-j}^{p^{e-i}})/K(L_{e-j-1}^{p^{e-i}})(L_{e-j+1}^{p^{e-i+1}})$ has exponent 1.

q.e.d.

<u>1.54. Example</u>. L/K has a lower tower generating set which is a minimal generating set, but which cannot be canonically ordered as in Proposition 1.52: Let

$L = K(z^{p^{-4}}, z^{p^{-4}}x^{p^{-2}} + y^{p^{-3}}, y^{p^{-2}})$, where $K = P(x,y,z)$, P is a perfect field and x,y,z are independent indeterminates over P. Then $\{T_{ij}\}$ is a lower tower generating system of L/K where: $T_{44} = \{z^{p^{-4}}, z^{p^{-4}}x^{p^{-2}} + y^{p^{-3}}\}$, $T_{34} = \{z^{p^{-3}}, z^{p^{-3}}x^{p^{-1}} + y^{p^{-2}}\}$, $T_{24} = \{z^{p^{-2}}\}$, $T_{14} = \{z^{p^{-1}}\}$, $T_{33} = T_{23} = T_{13} = \emptyset$, $T_{22} = \{y^{p^{-2}}\}$, $T_{12} = \{y^{p^{-1}}\}$ and $T_{11} = \emptyset$. Thus $K(L^{p^2}) = K(z^{p^{-2}}, z^{p^{-2}}x + y^{p^{-1}})$ $= K(z^{p^{-2}}, y^{p^{-1}}) \neq K(z^{p^{-2}}) = K(T_{13}, T_{24})$. $T_{11} \cup T_{22} \cup T_{33} \cup T_{44}$ is a minimal generating set of L/K.

If L/K has an exponent e, it is easily verified that if $\{T_{ij}\}$ is a lower tower system of L/K such that $T_{ii} = \emptyset$, $i = 1,\ldots,e-2$, then every lower tower generating set of L/K is a minimal generating set of L/K. Also if $|\bigcup_{j=1}^{e} T_{jj}| = 2$, then every lower tower generating set of L/K is a minimal generating set of L/K: By the comments above Example 1.45, $\bigcup_{j=1}^{e} T_{jj}$ contains a relative p-base M. If M is a singleton set, then since $L = K(M)$, $[L : K] = p^e$. However $[L : K] > p^e$ since $|\bigcup_{j=1}^{e} T_{jj}| = 2$. Thus $\bigcup_{j=1}^{e} T_{jj} = M$.

<u>1.55. Proposition.</u> Suppose L/K has an exponent e.

(a) If $e = 1$ or 2, every lower tower set of L/K is a minimal generating set of L/K and every lower tower set of L/K

can be canonically ordered as in Proposition 1.52.

(b) If $e = 3$ and there exists a lower tower set of L/K which is a minimal generating set of L/K, then every lower tower set of L/K can be canonically ordered as in Proposition 1.52.

(c) If L/K has a subbase, then every lower tower set is a subbase of L/K (whence a minimal generating set of L/K) and every lower tower set can be canonically ordered as in Proposition 1.52.

(d) If L/K has an equi-exponential subbase, then every lower tower set of L/K is an equi-exponential subbase, every lower tower set of L/K can be canonically ordered as in Proposition 1.52, and every relative p-base of L/K is a lower tower generating set of L/K.

Proof. (a) follows easily from the definition of a lower tower system. (c) is a consequence of Proposition 1.56 below. (d) follows from condition (c) and the invariance properties of canonical and lower tower generating systems. To prove (b), suppose $e = 3$ and let $T_{11} \cup T_{22} \cup T_{33}$ denote the given lower tower set. Then $L = K(T_{11}, T_{22}, T_{33})$ and it is easily verified that $K(L^{p^2}) = K(T_{13})$. Now T_{11} is a minimal generating set of $L_1/K(L_2^p)$. By hypothesis T_{11} is thus a minimal generating set of $L_1(L^p)/K(L^p)$. Hence

(*) L_1 and $K(L^p)$ are linearly disjoint over $K(L_2^p)$.

Now T_{23} is a minimal generating set of $L_1(L^p)/L_1$. Hence by (*), T_{23} is a minimal generating set of $K(L^p)/K(L_2^p)$. Now T_{12} is a minimal generating set of $K(L_2^p)/K(L_3^{p^2})$. Thus $K(L^p) = K(L_3^{p^2})(T_{12}, T_{23}) = K(T_{13}, T_{12}, T_{23}) = K(T_{12}, T_{23})$ since $T_{13} \subseteq T_{23}^p$. Therefore the desired result follows from Proposition 1.52. q.e.d.

D. Modular extensions. The existence of a subbase in L/K is a manifestation of a property that is invariantly described by: L and K^{p^i} are linearly disjoint over their intersection, $i = 1, 2, \ldots$. We now collect some useful facts about extensions with the latter property.

1.56. Proposition. If L/K has exponent e, then the following properties are equivalent:

(1) L/K has a subbase.

(2) There exists a canonical generating system such that $B_i^{p^i} \subseteq (L^{p^i} \cap K)(B_{i+1}^{p^i}, \ldots, B_e^{p^i})$, $i = 1, \ldots, e$.

(3) L^{p^i} and K are linearly disjoint over $L^{p^i} \cap K$, $i = 1, 2, \ldots$.

(4) Every canonical generating set is such that $B_i^{p^i} \subseteq (L^{p^i} \cap K)(B_{i+1}^{p^i}, \ldots, B_e^{p^i})$, $i = 1, \ldots, e$.

(5) For every lower tower set N, $h(x) = \ell(x)$ for all $x \in N$.

(6) For every lower tower set N, $\{x^{p^{h(x)}} \mid x \in N\}$ is a relative p-base of $(K^{p^{-1}} \cap L)/K$.

Proof. (1) implies (2): Any subbase can be canonically ordered by sorting on exponents. Hence condition (2) is clearly satisfied.

(2) implies (3): Let $\{B_1,\ldots,B_e\}$ be a canonical generating system satisfing condition (2). If we note that $L^{p^i} = (L^{p^{i-1}})^{p^i/p^{i-1}}$, then we find by induction that $L^{p^i} = (L^{p^i} \cap K)(B_{i+1}^{p^i},\ldots,B_e^{p^i})$, $i = 1,\ldots,e$. For a fixed i, for all $j > i$ and $b \in B_j$, we have $(b^{p^i})^{p^j/p^i} \in K(M_{j-1}^{p^i} - b^{p^i})$, where $M_j = B_{j+1} \cup \ldots \cup B_e$, $i = 0,1,\ldots,e$ and $K(M_e)$ means K. But $b^{p^{j-1}} \notin K(M_{j-1} - b)$ and $(b^{p^i})^{p^j/p^i} \in (L^{p^j} \cap K)(M_j^{p^j}) \subseteq (L^{p^i} \cap K)(M_{j-1}^{p^i} - b^{p^i})$, which implies condition (3).

(3) implies (4): Consider now an arbitrary canonical generating system $\{B_1,\ldots,B_e\}$. Then $K(L^{p^i}) = K(M_i^{p^i})$, $i = 1,\ldots,e$. From this and the linear disjointness of L^{p^i} and K, it follows that $L^{p^i} = (L^{p^i} \cap K)(M_i^{p^i})$, $i = 1,\ldots,e$. Thus $B_i^{p^i} \subseteq (L^{p^i} \cap K)(M_i^{p^i})$, $i = 1,\ldots,e$.

(4) implies (5): Since condition (4) implies condition (2) immediately, we have the equivalence of conditions (2), (3) and

(4). Hence we show that condition (3) implies condition (5). By the definition of lower tower system, it suffices to show that if a subset T of L_{i+1} is relatively p-independent over L_i, then T^p is relatively p-independent in L_i/L_{i-1}, $i = 1,\ldots,e$. This is equivalent to showing that the monomials $\prod_{x\in T} x^{pe_x}$, $0 \leq e_x < p$, all but a finite number of the e_x are zero, are linearly independent over L_{i-1}. Suppose there is a linear dependence among these monomials. If we raise both sides of such a dependence to the p^{i-1} power, we get a linear dependence over K among the monomials $\prod_{x\in T} x^{p^i e_x}$, $0 \leq e_x < p$, all but a finite number of the e_x are zero. Hence a linear dependence can be found with coefficients lying in $L^{p^i} \cap K$ since L^{p^i} and K are linearly disjoint. Taking the p^i root of such linear dependence gives a linear dependence over L_i among the monomials $\prod_{x\in T} x^{e_x}$, $0 \leq e_x < p$, all but a finite number of the e_x are zero. This contradicts the relative p-independence of T over L_i.

(5) implies (6): We always have that T_1 is a relative p-base of L_1/K by definition of lower tower system. Also, T_1 is a subset of the set given in (6) which is equal to that set if and only if (5) holds.

(6) implies (1): This follows from Proposition 1.44 in view of the comments made in (5) implies (6). q.e.d.

1.57. Definition. L/K is called modular if and only if L^{p^i} and K are linearly disjoint, $i = 1,2,\ldots$. When L/K is purely inseparable and has a subbase M, we shall hereafter call M a modular base.

1.58. Proposition. (1) There exists a maximal intermediate field K^* of L/K such that K^*/K is modular and (2) there exists a minimal intermediate field K^* of L/K such that L/K^* is modular.

Proof. (1) Let $S = \{K_j \mid K_j$ is an intermediate field of L/K and K_j/K is modular$\}$. Then S is partially ordered under set inclusion. Now $K \in S$ whence $S \neq \emptyset$. Let S' be any simply ordered subset of S. Let $K^* = \bigcup_{K_j \in S'} K_j$. Let i be a fixed but arbitrary positive integer. Let $X \subseteq K$ be a linear basis of K over $K^{*p^i} \cap K$. Suppose that $0 = \sum_{t=1}^{r} k_t^{*p^i} x_t$, there $x_1,\ldots,x_r \in X$ and $k_1^*,\ldots,k_r^* \in K^*$. Now there exists $K_j \in S'$ such that $k_1^*,\ldots,k_r^* \in K_j$. Since X is linearly independent over $K^{*p^i} \cap K$, X is linearly independent over the smaller field $K_j^{p^i} \cap K$. Since $K_j \in S$, X is linearly independent over $K_j^{p^i}$, whence $k_t^{*p^i} = 0 (t = 1,\ldots,r)$. Thus, K^{*p^i} and K are linearly disjoint. Hence, $K^* \in S$ whence S has a maximal element.

(2) Let $S = \{K_j \mid K_j$ is an intermediate field of L/K and L/K_j is modular$\}$. Then S is partially ordered under set containment. Now $L \in S$ whence $S \neq \emptyset$. Let S' be any simply

ordered subset of S. Let $K^* = \bigcap_{K_j \in S'} K_j$. Let i be a fixed but arbitrary positive integer. Let $X \subseteq L^{p^i}$ be a linear basis of L^{p^i} over $L^{p^i} \cap K^*$. Suppose that $0 = \sum_{t=1}^{r} k_t^* x_t$, where $k_t^* \in K^*$ and $x_t \in X$, $t = 1,\ldots,r$. Clearly x_1 is linearly independent over $L^{p^i} \cap K_j$ for any $K_j \in S'$. Make the induction hypothesis that $\{x_1,\ldots,x_m\}$, $1 \leq m < r$, is linearly independent over $L^{p^i} \cap K_{j_0}$ for some $K_{j_0} \in S'$. Let $S_0' = \{K_j \mid K_j \in S',$ $K_j \subseteq K_{j_0}\}$. Then $\{x_1,\ldots,x_m\}$ is clearly linearly independent over $L^{p^i} \cap K_j$ for all $K_j \in S_0'$. If x_{m+1} is in the linear span of $\{x_1,\ldots,x_m\}$ over $L^{p^i} \cap K_j$ for all $K_j \in S_0'$, then x_{m+1} is a linear combination of $x_1,\ldots,x_m$ over $L^{p^i} \cap K_j$ for all $K_j \in S_0'$. Equating these linear combinations, we find that the coefficients all lie in $\bigcap_{K_j \in S_0'} (L^{p^i} \cap K_j) = L^{p^i} \cap K^*$. However, this contradicts the linear independence of $\{x_1,\ldots,x_r\}$ over $L^{p^i} \cap K^*$. Hence, there exists $K_{j_1} \in S_0' \subseteq S'$ such that $\{x_1,\ldots,x_{m+1}\}$ is linearly independent over $L^{p^i} \cap K_{j_1}$. Therefore, by induction, there exists $K_j \in S'$ such that $\{x_1,\ldots,x_r\}$ is linearly independent over $L^{p^i} \cap K_j$. Since $K_j \in S$, $\{x_1,\ldots,x_r\}$ remains linearly independent over K_j. Since $k_1^*,\ldots,k_r^* \in K^* \subseteq K_j$, $k_1^* = \ldots = k_r^* = 0$. Thus, X is linearly independent over K^*. Therefore, $K^* \in S$ whence S has a minimal element. q.e.d.

Since the existence of a subbase for a purely inseparable extension L/K is equivalent to the modularity of L/K in the bounded exponent case, Proposition 1.58 proves the existence of a maximal intermediate field with a modular base and a minimal intermediate field over which L has a modular base.

1.59. Example. Not every purely inseparable extension L/K is modular: Let $L = K(z^{p^{-2}}, z^{p^{-2}}x^{p^{-1}} + y^{p^{-1}})$ where $K = P(x,y,z)$, P is a perfect field and x,y,z are independent indeterminates over P. Then $\{T_{ij}\}$ is a lower tower system of L/K where $T_{22} = \{z^{p^{-2}}, z^{p^{-2}}x^{p^{-1}} + y^{p^{-1}}\}$, $T_{12} = \{z^{p^{-1}}\}$ and $T_{11} = \emptyset$. Since $\ell(z^{p^{-2}}x^{p^{-1}} + y^{p^{-1}}) \neq h(z^{p^{-2}}x^{p^{-1}} + y^{p^{-1}})$, L/K is not modular by Proposition 1.56.

The remainder of this section is devoted to the concept of modular closure as developed in [53].

1.60. Lemma. Let $F \supseteq L \supseteq K$ be fields.

(a) L and $K(F^{p^i})$ are linearly disjoint over $K(L \cap F^{p^i})$, $i = 1,2,\ldots,$ and F/K is modular if and only if $L \cap F^{p^i}$ and K are linearly disjoint for $i = 1,2,\ldots,$ and F/L is modular.

(b) L^{p^i} and $F^{p^i} \cap K$ are linearly disjoint, $i = 1,2,\ldots,$ and F/K is modular if and only if F^{p^i} and $K(L^{p^i})$ are linearly disjoint over $L^{p^i}(K \cap F^{p^i})$, $i = 1,2,\ldots,$ and L/K is modular.

(c) L and $K(F^{p^i})$ are linearly disjoint over $K(L \cap F^{p^i})$, $i = 1,2,\ldots$, L^{p^i} and $F^{p^i} \cap K$ are linearly disjoint, $i = 1,2,\ldots$, and F/K is modular if and only if $L \cap F^{p^i}$ and $K(L^{p^i})$ are linearly disjoint over $L^{p^i}(K \cap F^{p^i})$, $i = 1,2,\ldots$, F/L and L/K are modular.

Proof. This follows from the application of a well-known property of linear disjointness (see Lemma in [24, p.162]) to the following diagram:

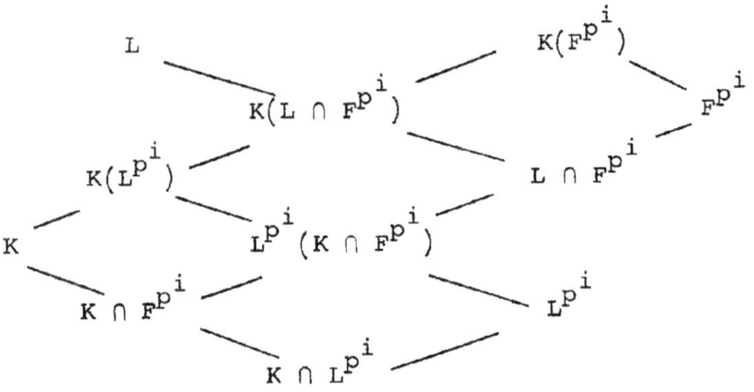

q.e.d.

<u>1.61. Lemma</u>. Let $F \supseteq L \supseteq K$ be fields.

(a) If L/K is separable, then L/K is modular.

(b) If L/K is separable algebraic, then F/K is modular implies F/L is modular.

(c) If F/L is separable, then F/K is modular if and only if L/K is modular.

Proof. (a) Since L and $K^{p^{-i}}$ are linearly disjoint over K, so are L^{p^i} and K over $K^{p^i} = L^{p^i} \cap K$.

(b) Since L/K is separable algebraic, $L = K(L^{p^i}) \subseteq K(F^{p^i})$ and $L^{p^i} \subseteq F^{p^i}$, the result follows from the diagram

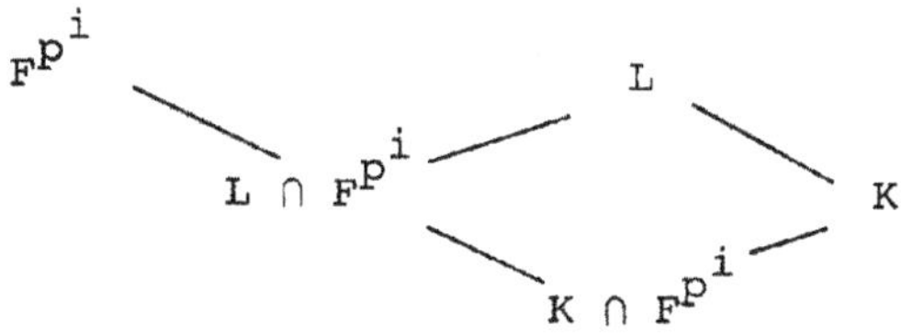

(Note: L/K relative perfect is sufficient.)

(c) Since F/L is separable, $L \cap F^{p^i} = L^{p^i}$ and $K \cap F^{p^i} = K \cap L^{p^i}$. The result follows from the diagram

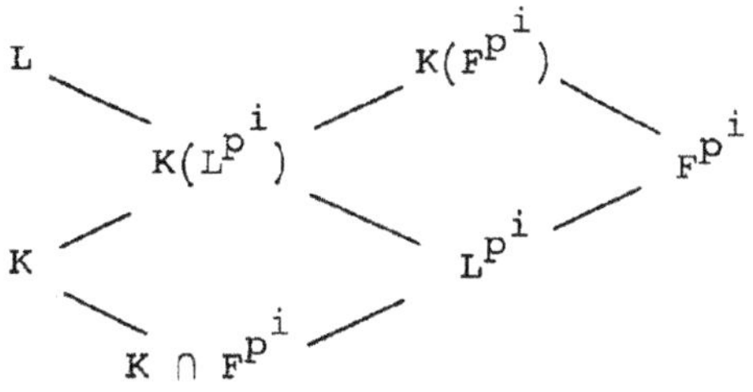

q.e.d.

<u>1.62. Lemma</u>. (a) For each pair L and J (fields), there is a unique minimal extension, F/L which is linearly disjoint from J.

(b) $F = L[S]$ for a subset S of J.

(c) If $[L : L \cap J] < \infty$, then S may be chosen to be finite.

Proof. Let $\{x_i\}$ denote a linear basis for $L/L \cap J$, and let $\{x_j\}$ denote a maximal subset of $\{x_i\}$ which is linearly independent over J in $L(J)$. Then $\{x_j\}$ is a linear basis for $L(J)/J$. Set $\{x_t\} = \{x_i\} - \{x_j\}$. For each x_t, we have

$$(*) \qquad x_t = \sum_j a_{tj} x_j$$

for unique $\{a_{tj}\} \subseteq J$. Let $S = \{a_{tj}\}$ and $F = L[S]$. Notice, S is finite when $L/L \cap J$ is finite. Since $\{x_j\}$ is linearly independent over J, $\{x_j\}$ is linearly independent over $F \cap J$. By condition (*), L lies in the span of $\{x_j\}$ over $F \cap J$ (as does $S \subseteq F \cap J$). Thus $F = L[S]$ lies within this span. Hence $\{x_j\}$ is a basis of $F(J)/J$, so that F and J are linearly disjoint. Now suppose M is a field such that M and J are linearly disjoint and $M \supseteq L$. Then $\{x_j\} \subseteq M$ and $\{x_j\}$ is linearly independent over $M \cap J$. For each x_t, we have (*) with the $\{a_{tj}\}$ unique. By the linear disjointness of M and J over $M \cap J$ these relations must occur over $M \cap J$. Hence $\{a_{tj}\} \subseteq M \cap J \subseteq M$. Thus $M \supseteq L[S] = F$. q.e.d.

<u>1.63 Proposition</u>. Let L/K have exponent e. There exists a unique minimal field F, $F \supseteq L$, such that:

(a) F/K is modular.

(b) F/K is purely inseparable.

(c) F/K has exponent e.

(d) If L/K is finite, then F/K is finite.

Proof. F is constructed by descending induction. Suppose we have constructed $F_m \supseteq L$ satisfying:

(1) $F_m^{p^s}$ and K are linearly disjoint, $s = m, m+1, \ldots,$

(2) F_m is the unique minimal extension of L having property (a),

(3) F_m/K is purely inseparable,

(4) F_m/K has exponent e,

(5) if $[L:K] < \infty$, then $[F_m : K] < \infty$.

Note that the induction begins with $F_e = L$. By Lemma 1.62, there exists a unique minimal field $M \supseteq F_m^{p^{m-1}}$, where M and K are linearly disjoint. Let $F_{m-1} = M^{p^{-(m-1)}}$. Then $F_{m-1}^{p^{m-1}}$ and K are linearly disjoint. By Lemma 1.62, $M = F_m^{p^{m-1}}[S]$ for a subset S of K. Thus for $s = m, m+1, \ldots, F_{m-1}^{p^s} = M^{p^{s-m+1}} = F_m^{p^s}[S^{p^{s-m+1}}]$ and $F_{m-1}^{p^s} = F_m^{p^s}(F_{m-1}^{p^s} \cap K)$. Because $F_m^{p^s}$ and K are linearly disjoint for $s \geq m$, we have that $F_{m-1}^{p^s}$ and K are linearly disjoint. Thus F_{m-1} satisfies (1). Also, since $F_{m-1}^{p^e} = F_m^{p^e}(F_{m-1}^{p^e} \cap K)$, it follows that F_m satisfies (3) and (4). (Here we have assumed $e \geq m$ which is possible since induction begins with F_e.) If $[L:K] < \infty$, then $[F_m : K] < \infty$ by (5) and $[F_m^{p^{m-1}} : F_m^{p^{m-1}} \cap K] < \infty$. By Lemma 1.62, we may assume S is a finite subset of K. Now $F_{m-1} = M^{p^{-(m-1)}} = (F^{p^{m-1}}[S])^{p^{-(m-1)}} = F_m[S^{p^{-(m-1)}}]$. This is a finite extension of K. It remains to verify (2). Suppose N

is a field such that $N \supseteq L$ and N satisfies (1). Then $F_m \subseteq N$ since N^{p^m} and K are linearly disjoint. Hence $F_m^{p^{m-1}} \subseteq N^{p^{m-1}}$ which is linearly disjoint from K. Hence by Lemma 1.62, $M \subseteq N^{p^{m-1}}$. Thus $F_{m-1} \subseteq N$. The induction leads to F_1, the desired field. q.e.d.

<u>1.64. Definition</u>. If a field F satisfies the conditions of Proposition 1.63, then F is called the modular closure of L/K.

<u>1.65. Corollary</u>. For each finite normal extension L/K, there is a unique minimal extension F over L which is modular over K. F is a finite purely inseparable extension of L.

Proof. Let S be the maximal separable intermediate field of L/K. Then L/S is purely inseparable. By Proposition 1.63, let F be the modular closure of L/S. Then F is a finite purely inseparable extension of L so that F is still normal over K and S is the maximal separable intermediate field of F/K. As is well-known, this situation implies $F = M \otimes_K S$, where M is the maximal purely inseparable intermediate field of F/K. Since S/K is finite separable $S = K(s) = K(s^{p^n})$ for some $s \in S$ and all positive integers n. Thus if $[S : K] = t$, we have that $\{1, s^{p^n}, s^{2p^n}, \ldots, s^{(t-1)p^n}\}$ is a linear basis of S/K and any element a of F can be written uniquely as $a = \sum_{i=0}^{t-1} m_i s^i$, where $\{m_i\} \subseteq M$. Then

$a^{p^n} = \sum_{i=0}^{t-1} m_i^{p^n} s^{ip^n}$ and since $\{1, s^{p^n}, \ldots, s^{(t-1)p^n}\}$ is a linear basis of S/K we have that a^{p^n} belonging to S implies $\{m_i^{p^n}\} \subseteq K$. Thus $F^{p^n} \cap S = S^{p^n}(M^{p^n} \cap K)$. Also, a^{p^n} belonging to K implies $0 = m_1 = \ldots = m_{t-1}$, so $a = s_0 \in M$. Thus $F^{p^n} \cap K = M^{p^n} \cap K$. Hence $F^{p^n} \cap S = S^{p^n}(F^{p^n} \cap K)$. Since S/K is separable we have $K(S^{p^n}) = S$. Since F/S and S/K are modular, it follows by the following diagram that F/K is modular:

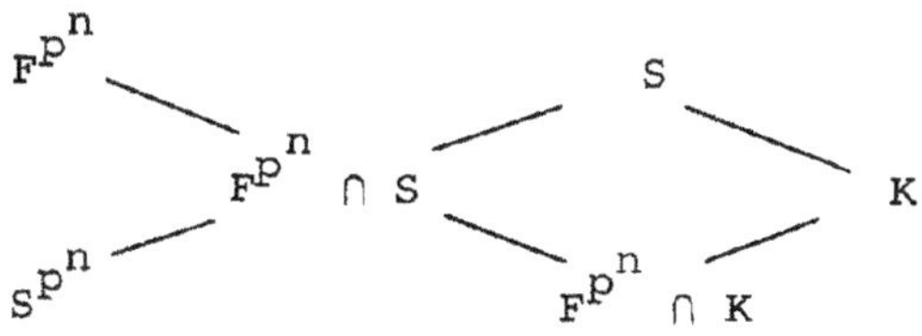

The additional properties of F follow from parts (b) and (d) of Proposition. 1.63. q.e.d.

The field F in Corollary 1.65 is also called the modular closure of L/K.

<u>1.66. Corollary</u>. For each finite extension L/K, there is a unique minimal extension M over L which is normal and modular over K. M is finite over K.

Proof. Let F be the normal closure of L/K and M the modular closure of F/K. Since M is a purely inseparable extension of F, it is a normal extension of K. The minimality of M is clear. F is a finite extension of K. Thus by the previous Corollary, M/K is finite. q.e.d.

Suppose M/K is a finite normal modular extension. Then $M = L \otimes_K F$ where L/K is Galois and F/K is purely inseparable, by Theorem 13, [24, p.52]. Also M/F is separable and F/K is modular. Hence $F = K(x_1) \otimes_K \cdots \otimes_K K(x_t)$ and $M = K(x_o) \otimes_K \cdots \otimes_K K(x_t)$, where $L = K(x_o)$.

E. Extension exponents. We now look more closely at the role of the base field in determining the properties of an extension.

For a subset C of L, we let $C^{p^{-\infty}}$ denote the set $\{c^{p^{-i}} \mid c \in C,\ i = 1,2,\ldots\}$, and let $c^{p^{-\infty}}$ denote $\{c\}^{p^{-\infty}}$.

1.67. Definition. L/K is called purely infinite if and only if there exists a set C in $K - K^p$ such that $L = K(C^{p^{-\infty}})$.

1.68. Proposition. Suppose L/K is purely infinite.

(a) If $L = K(C^{p^{-\infty}})$ where $K = K^p(C)$, then L is the smallest perfect extension of K.

(b) $L = K(P)$ where P is the maximal perfect subfield of L.

(c) L/K is relatively perfect and every p-base of K contains a p-base of L. If G is a subset of K such that G is p-independent in L and $L^p(G) = L^p(K)$, then G is a p-base of L.

Proof. (a) $L^p = K^p(C^{p^{-\infty}}) = K^p(C)(C^{p^{-\infty}}) = K(C^{p^{-\infty}}) = L$ whence L is perfect. Let P' be a perfect field such that $K \subseteq P' \subseteq L$. Then clearly $C^{p^{-\infty}} \subseteq P'$ so $P' = L$.

(b) $L \supseteq K(P) = K(\bigcap_{i=1}^{\infty} L^{p^i}) = K(\bigcap_{i=1}^{\infty} K^{p^i}(C^{p^{-\infty}})) \supseteq K(C^{p^{-\infty}}) = L$

where C is a subset of $K - K^p$.

(c) $K(L^p) = K(K^pP^p) = K(P) = L$. Let D be a p-base of K. Then $L^p(D) = K^pP^p(D) = K(P) = L$. Consider the set G. Let M be a relative p-base of L/K. Then $G \cup M$ is a p-base of L. Since L/K is relatively perfect, $M = \emptyset$. q.e.d.

<u>1.69. Proposition</u>. Suppose L/K is purely infinite. Let P denote the maximal perfect subfield of L. Then the following conditions are equivalent.

(1) K and P are linearly disjoint.

(2) There exists a p-independent set C in K such that $L = K(C^{p^{-\infty}})$.

(3) L/K is modular.

Proof. (1) implies (2): Take G as in part (c) of Proposition 1.68. Then G is a p-base of L. Since K and P are linearly disjoint and $(K \cap P)(K^p)(G) \subseteq K$, $(K \cap P)(K^p)(G)$ and P are linearly disjoint over $K \cap P$. Since $L^p = K^p(P)$ and G is a p-base of L, $L = (K \cap P)(K^p)(G)(P)$. Since, also, $L = K(P)$, we have by the linear disjointness of K, P and $(K \cap P)(K^p)(G)$, P over $K \cap P$ that $K = (K \cap P)(K^p)(G)$.

Thus $K \cap P$ contains a set C such that $G \cup C$ is a p-base of K. It follows that G is a p-base of $K(C^{p^{-\infty}})$. Hence $L/K(C^{p^{-\infty}})$ preserves p-independence. Hence, by Proposition 1.13, $L/K(C^{p^{-\infty}})$ is separable and algebraic. Since $L/K(C^{p^{-\infty}})$ is also purely inseparable, $L = K(C^{p^{-\infty}})$.

(2) implies (3): Let $G \subseteq K$ be such that $G \cup C$ is a p-base of K. Since $L = K(C^{p^{-\infty}})$, G is a p-base of L. Now $K \supseteq (L^{p^i} \cap K)(G) \supseteq K^{p^i}(C,G) = K$, $i = 1,2,\ldots$. Thus $K = (L^{p^i} \cap K)(G)$, $i = 1,2,\ldots$. Since G is a p-base of L, K and L^{p^i} are linearly disjoint $(i = 1,2,\ldots)$.

(3) implies (1): Let X be a maximal subset of P which is linearly independent over K. Then X is certainly linearly independent over the smaller fields $K \cap L^{p^i}$, $i = 1,2,\ldots$. Moreover because K and L^{p^i} are linearly disjoint and $P \subseteq L^{p^i}$, X remains maximally linearly independent over $K \cap L^{p^i}$ $(i = 1,2,\ldots)$. Thus X is a linear basis of $L^{p^i}/K \cap L^{p^i}$, $i = 1,2,\ldots$. For any element b of P, $b = \sum k_{i_0} x_{i_0} = \sum k_{i_1} x_{i_1} = \ldots$ where $k_{i_j} \in K \cap L^{p^i}$ and $x_{i_j} \in X$, $j = 0,1,\ldots$. Hence the coefficients k_{i_j} of these linear equations are in $\bigcap_{i=1}^{\infty} (K \cap L^{p^i}) = K \cap (\bigcap_{i=1}^{\infty} L^{p^i}) = K \cap P$. Thus X is a linear basis

of $P/K \cap P$. Condition (1) holds since X is linearly independent over K. q.e.d.

1.70. Definition. Let L/K be purely inseparable. For $c \in K - K^p$, if there exists a non-negative integer n such that $c^{p^{-n}} \in L - L^p$, then n is called the extension exponent of c with respect to L.

1.71. Definition. An element a in $K - K^p$ is called U-dependent on a subset C of $K - K^p$ if and only if there exists a finite set $D \subseteq C$ such that $K(a^{p^{-\infty}}) \subseteq K(D^{p^{-\infty}})$.

1.72. Proposition. For $a,b \in K - K^p$, the U-dependence of a on b is equivalent to $K^{p^i}(a) = K^{p^i}(b)$, $i = 1,2,\ldots$.

Proof. That these equations imply U-dependence is trivial. Conversely suppose that a is U-dependent on b. Let i be any positive integer. Then $a^{p^{-i}} \in K(b^{p^{-j_i}})$ for some positive integer j_i. Hence $K(a^{p^{-i}}) = K(b^{p^{-j}})$ for some positive integer j. Thus $i = j$ by a degree argument. Therefore $K^{p^i}(a) = K^{p^i}(b)$. q.e.d.

1.73. Proposition. U-dependence satisfies the dependence axioms.

Proof. We give the proof of the exchange property only. Let C be a finite subset of $K - K^p$ and set $K(C^{p^{-\infty}}) = L$.

If b is U-dependent on $\{a\} \cup C$ but U-independent of C, then a is U-independent of C. Thus a has an extension exponent i with respect to L. Likewise b has an extension exponent j with respect to L. Set $a' = a^{p^{-i}}$ and $b' = b^{p^{-j}}$. The U-dependence of b on $\{a\} \cup C$ yields the U-dependence of b' on a' in L. That is, $K(\{b\}^{p^{-\infty}}) \subseteq K(C^{p^{-\infty}})(\{a\}^{p^{-\infty}})$ implies $K(\{b'\}^{p^{-\infty}}) \subseteq K(C^{p^{-\infty}})(\{a'\}^{p^{-\infty}})$ which implies $K(C^{p^{-\infty}})(\{b'\}^{p^{-\infty}}) \subseteq K(C^{p^{-\infty}})(\{a'\}^{p^{-\infty}})$. This latter inclusion implies b' is U-dependent on a' in L. Thus, by Proposition 1.72, $L^{p^i}(b') = L^{p^i}(a')$ for all positive integers i. Hence a' is U-dependent on b' in L. Therefore $K(\{a'\}^{p^{-\infty}}) \subseteq K(C^{p^{-\infty}})(\{b'\}^{p^{-\infty}})$ whence $K(\{a\}^{p^{-\infty}}) \subseteq K(C^{p^{-\infty}})(\{b\}^{p^{-\infty}})$. Thus we have the interchangeability of a and b. q.e.d.

1.74. Example. It is essential that the finite subset D occurs instead of C in Definition 1.71: Let $K = P(x, x_1, x_2, \ldots)$ where P is a perfect field and $x, x_1, x_2, \ldots$ are independent indeterminates over P. For $C = \{xx_1^p, xx_2^{p^2}, \ldots\}$, we have $K(\{x\}^{p^{-\infty}}) \subseteq K(C^{p^{-\infty}})$. However for $D = \{xx_1^p, \ldots, xx_n^{p^n}\}$, $K(\{x\}^{p^{-\infty}}) \not\subseteq K(D^{p^{-\infty}})$.

Pickert used the theory of U-dependence to compare the imperfection degrees of L and K in an algebraic extension L/K. He showed that the imperfection degree of L is not larger than that of K, and that when L/K has an exponent the degrees are equal. In connection with these facts, we remark that a purely inseparable L/K which has a modular base will have the same imperfection degree as K because the base arises from a p-independent subset of K. See [43, p.132]. Hence every intermediate field of L/K has the same imperfection degree.

1.75. Definition. A purely inseparable extension L/K is called semi-finite if and only if every element of $K - K^p$ has an extension exponent with respect to L.

1.76. Proposition. Let L/K be purely inseparable. Then there exists a unique maximal purely infinite intermediate field of L/K and this intermediate field has the property that it is the only purely infinite intermediate field over which L is semi-finite.

Proof. Let P be the maximal perfect subfield of L and $C \subseteq K - K^p$ the set of all elements without extension exponent with respect to L. Then $C = K \cap P - K^p$, whence $K(C^{p^{-\infty}}) = K(P)$. Clearly $K(C^{p^{-\infty}})$ contains all other purely infinite intermediate fields of L/K. Every element $b \in K(C^{p^{-\infty}}) - K^p(C^{p^{-\infty}})$ has an extension exponent with respect to L, for otherwise $b^{p^e} \in C$ where e is the exponent of b over K whence $b \in K^p(C^{p^{-\infty}})$. Hence $L/K(C^{p^{-\infty}})$ is semi-finite. If $K(D^{p^{-\infty}}) \subset K(C^{p^{-\infty}})$, there exists a least one element c in C with extension exponent with respect to $K(D^{p^{-\infty}})$. Then $b = c^{p^{-e}} \in K(D^{p^{-\infty}}) - K^p(D^{p^{-\infty}})$.

But since c possesses no extension exponent with respect to L, b also has none. That is, $L/K(D^{p^{-\infty}})$ is not semi-finite. q.e.d.

1.77. Proposition. Let L/K be purely inseparable and let P and Q denote the maximal perfect subfields of L and K, respectively. Then L/K is semi-finite if and only if $P = Q$.

Proof. By the proof of Proposition 1.76, we have that $K(P)$ is the maximal purely infinite intermediate field of L/K. Thus, by Proposition 1.76, L/K is semi-finite if and only if $P \subseteq K$. Clearly $P \subseteq K$ if and only if $P = Q$. q.e.d.

1.78. Definition. Let K be a field. A subset C of $K - K^p$ is called U'-independent in K if and only if for all $c \in C$ there exists a positive integer i_c such that $c^{p^{-i_c}} \notin K((C - c)^{p^{-\infty}})$.

For all integers $i \geq i_c$, $c^{p^{-i}} \notin K((C - c)^{p^{-\infty}})$ when C is U'-independent in K. The smallest such positive integer i_c is called the extension exponent of c with respect to C. The set $M = \{c^{p^{-i}} \mid c \in C,\ i$ is any fixed integer $\geq i_c\}$ is called an extension set of C in L.

An element a in $K - K^p$ is thus U'-dependent on a subset C of $K - K^p$ if and only if $K(a^{p^{-\infty}}) \subseteq K(C^{p^{-\infty}})$.

1.79. Definition. Let L/K be purely inseparable. A subset C of $K - K^p$ is called relatively U'-independent in L/K if and only if for all $c \in C$, c has a positive extension exponent $i = i_c$ in L and $c^{p^{-i}} \notin K(L^p)((C - c)^{p^{-\infty}})$. $M = \{c^{p^{-i}} \mid c \in C,\ i = i_c\}$ is called the extension set of C in L.

1.80. Proposition. Let L/K be purely inseparable. If C is a maximal relatively U'-independent set in L/K and M is the extension set of C in L, then M is a relative p-base of L/K. Conversely if M is a relative p-base of L/K, then $C = \{m^{p^i} \mid m \in M,$ i is the exponent of m over $K\}$ is a maximal relatively U'-independent set in L/K.

Proof. It suffices to show that C is relatively U'-independent in L/K if and only if M is relatively p-independent in L/K. Given C, $c^{p^{-i}} \notin K(L^p)((C - c)^{p^{-\infty}})$ implies $c^{p^{-i}}$ is not in the smaller field $K(L^p)(M - c^{p^{-i}})$, whence M is relatively p-independent in L/K. Given M, if there exists $c \in C$ such that $c^{p^{-i}} \in K(L^p)((C - c)^{p^{-\infty}})$, then there exists a positive integer j such that $c^{p^{-i}} \in K(L^p)((C - c)^{p^{-j}}) \subseteq K(L^p)((M - c^{p^{-i}})^{p^{-j}})$. Thus $((M - c^{p^{-i}})^{p^{-j}}) \cup \{c^{p^{-i}}\}$ is not relatively p-independent in $L((M - c^{p^{-i}})^{p^{-j}})/K$ while M is relatively p-independent in L/K. However this is impossible by Proposition 1.7. Thus C is relatively U'-independent in L/K.

q.e.d.

1.81. Corollary. Let L/K be purely inseparable. B is a modular base of L/K if and only if K and L^p are linearly disjoint and B is an extension set of a maximal relatively U'-independent set C in L/K such that $K \cap L^p = K^p(C)$.

Proof. Suppose B is a modular base of L/K. Then K and L^{p^i}, $i = 1,2,\ldots,$ are linearly disjoint by Proposition 1.23. Now B is a relative p-base of L/K whence there exists a subset G of K such that G is p-independent in K, $L^p(K) = L^p(G)$, and $G \cup B$ is a p-base of L. Since K and L^p are linearly disjoint, $K = (K \cap L^p)(G)$. By Proposition 1.22 $G \cup C$ is a p-base of K where C is the relatively U'-independent set in L/K for which B is the extension set. Thus from $K = K^p(G,C)$ and $K = (K \cap L^p)(G)$, it follows that $K \cap L^p = K^p(C)$. Conversely suppose K and L^p are linearly disjoint and B, C are given. By Proposition 1.80 B is a relative p-base of L/K. Hence as above there exists a subset G of K such that $G \cup B$ is a p-base of L and $K = (K \cap L^p)(G)$ Since $K \cap L^p = K^p(C)$, $K = K^p(G,C)$. Thus B is a modular base of L/K by Proposition 1.22. q.e.d.

<u>1.82. Corollary</u>. Let K be a field. If C is an U'-independent set in K and M is an extension set of C, then M is a minimal generating set of L/K where $L = K(M)$. Conversely suppose M is a minimal generating set of a purely inseparable extension L/K. Then $C = \{m^{p^i} \mid m \in M,$ i is the exponent of m over $K\}$ is U'-independent in K and M is an extension set of C.

Proof. Suppose C is U'-independent in K. Then for any $c \in C$, $c^{p^{-i}} \notin K((C - c)^{p^{-\infty}})$ whence $c^{p^{-i}}$ is not in the smaller field $K(M - C^{p^{-i}})$. Thus M is a minimal generating set of L/K. Conversely suppose M is a minimal generating set of L/K. Then M is a relative p-base of L/K. By Proposition 1.80, C is relatively U'-independent in L/K whence clearly U'-independent in K. q.e.d.

1.83. Remark. Let L/K be purely inseparable. Consider the properties: (1) L/K is semi-finite, (2) L/K is not semi-finite, (3) L/K has a minimal generating set and (4) L/K is of type R. None implies the other except for (4) implies (3): Example 1.26 (a) shows that (4) $\not\Rightarrow$ (1) and Example 1.26 (b) shows that (4) $\not\Rightarrow$ (2). The extension $K(L^p)/K$ in Example 1.26 (b) shows that (1) $\not\Rightarrow$ (3). Letting L be perfect gives us an example showing that (2) $\not\Rightarrow$ (3). Let $K = P(x_1, x_2, \ldots)$ and $L = K(x_1^{p^{-1}}, x_2^{p^{-2}}, \ldots)$ where P is perfect and $x_1, x_2, \ldots$ are independent indeterminates over P. Then $\{x_1^{p^{-1}}, x_2^{p^{-2}}, \ldots\}$ is a minimal generating set of L/K, but $\{x_1^{p^{-1}} x_2^{p^{-1}}, x_2^{p^{-2}} x_3^{p^{-2}}, \ldots\}$ is a relative p-base of L/K which is not a minimal generating set of L/K. Thus (3) $\not\Rightarrow$ (4). The remaining implications are trivial.

1.84. Proposition. Let Q be the maximal perfect subfield of K and C a subset of $K - K^p$. Each statement in the following list implies the succeeding one.

(1) C is p-independent in K.

(2) C is U'-independent in K.

(3) C is U-independent in K.

(4) C is algebraically independent over Q.

Proof. (1) implies (2): If C is p-independent in K, then $C^{p^{-1}}$ is a minimal generating set of $K(C^{p^{-1}})/K$. Hence C is U'-independent in K by Corollary 1.82.

(2) implies (3): This implication follows immediately from the definitions.

(3) implies (4): Suppose there exists a finite subset $\{c, c_1, \ldots, c_n\}$ of C such that c is algebraic over $Q(c_1, \ldots, c_n)$. Then c is separable over $Q(c_1, \ldots, c_n)(\{c_1, \ldots, c_n\}^{p^{-\infty}})$ since the latter field is perfect. Thus $Q(c, c_1, \ldots, c_n)(\{c_1, \ldots, c_n\}^{p^{-\infty}})$ is also perfect. Hence $c^{p^{-i}} \in Q(c, c_1, \ldots, c_n)(\{c_1, \ldots, c_n\}^{p^{-\infty}})$, $i = 1, 2, \ldots$. Thus C is not U -independent in K. q.e.d.

<u>1.85. Remark</u>. Consider the conditions (1), (2), (3) and (4) in Proposition 1.84. Then (4) $\not\Rightarrow$ (3) $\not\Rightarrow$ (2) $\not\Rightarrow$ (1): Examples are easily constructed showing that (3) $\not\Rightarrow$ (2) $\not\Rightarrow$ (1). We show that (4) $\not\Rightarrow$ (3) even if C is a finite algebracically independent set over Q. Let Q' be a perfect field and $z_1, z_2, \ldots$ independent indeterminates over Q'. Consider $Q'(Z^{p^{-\infty}})$ where $Z = \{z_1, z_2, \ldots\}$. Let y and u_o be algebraically independent over $Q'(Z^{p^{-\infty}})$. Define recursively $u_n = y^{p^{n-1}} + z_n u_{n-1}, n = 1, 2, \ldots$. Set $K = Q'(Z^{p^{-\infty}}, y, u_o, u_1^{p^{-1}}, u_2^{p^{-2}}, \ldots, u_n^{p^{-n}}, \ldots)$. Set $C = \{y, u_o\}$. Then $u_o^{p^{-i}} \in K(y^{p^{-\infty}})$, $i = 1, 2, \ldots$. Thus C is not U-independent in K. However, as shown in [28], $Q = Q'(Z^{p^{-\infty}})$ is the maximal perfect subfield of K and C is algebraically independent over Q.

Reference Notes for Chapter I

The early definition and use of p-independence is attributed to Teichmuller [56]. The result 1.6 was noted by Rygg [48]. The basic idea in 1.7 is from Pickert [42]. The part of 1.8 concerning the monotone property was shown by Pickert [43]. Connections between separability and preservation of p-independence such as those in 1.13 were made by Teichmuller [56] and investigated by many others including MacLane [27], [28], and Dieudonne [9]. The result 1.11 appears in MacLane [28] and Dieudonne [9]. The definition of subbase, 1.21, and its connection to higher derivations occurs in Weisfeld [57]. The connection between subbase and linear disjointness was made independently in [14], [31], and by Sweedler in [53]. The canonical generator theory was started in [18], [31] and is based on Pickert's finite degree theory [43], [44]. The generating system in 1.38 is an outgrowth of one used by Sweedler [53] to analyze fields with exponent and subbase, and the definition of modular in 1.57 as well as the results in 1.60-1.66 are also due to Sweedler [53]. Most of the definitions and results 1.67, 1.68, 1.70-1.77 are due to Pickert [43]. The concept of type R was introduced and applied in [16], [33], [35], [37]; the theory of tower systems and related systems was initiated in [15], [16]; The concept of U'-independence appears here for the first time.

II. Intermediate Fields

A. Lattice invariants. Our first concern is with the lattice invariants arising from our special generating systems.

2.1. Definition. By the lattice $\mathcal{L}$ of an extension L/K we mean the set of all intermediate fields of L/K with intersection taken as meet (glb) and with composite taken as join (lub). A linearly ordered subset of $\mathcal{L}$ is called a chain.

2.2. Proposition. Let L/K be purely inseparable and let $\mathcal{L}$ be a chain. Then every proper intermediate field of L/K is simple. The number of proper intermediate fields is finite when L/K has an exponent, and is countably infinite when L/K has no exponent.

Proof. If $L \neq K(L^p)$, then a relative p-base of L/K must consist of exactly one element, say a. Hence $L = K(L^p)(a) \subseteq K(a) = L$, so $L = K(a)$. On the other hand, if $L = K(L^p) \neq K$ and if G is a proper intermediate field, there exists an element say a, such that $a \in L - G$. Hence $G \subset K(a)$, that is G/K is simple. Finally, we note that L is the union of the $K^{p^{-i}} \cap L$, and that $|S(K^{p^{-i}} \cap L)| < \infty$. Since $S(L)$ is thus a denumerable sum of finite sets, it is denumerable. (Note that the only proper intermediate fields must in fact be those of the form $K^{p^{-i}} \cap L$.) q.e.d.

2.3. Definition. When the cardinality of a relative p-base of L/K is m (that is, the relative imperfection degree is m), we call L/K an m-fold extension.

2.4. Proposition. Let L/K be a purely inseparable finite degree m-fold extension. Then $\mathcal{L}$ is a modular lattice if and only if $m \leq 2$. If $m > 2$, then $\mathcal{L}$ is an upper semi-modular lattice.

Proof. Define the function d from the set of all intermediate fields of L/K into the non-negative integers as follows: If L' is an intermediate field of L/K, then $p^{d(L')} = [L' : K]$. Let L' and L'' be intermediate fields of L/K. Then $d(L'L'') - d(L') \leq d(L'') - d(L' \cap L'')$ always holds (upper semi-modularity). Suppose $m \leq 2$. To prove that $\mathcal{L}$ is modular, we must show that

$$(*) \qquad d(L') + d(L'') = d(L' \cap L'') + d(L'L'').$$

If $m = 1$, then L/K is a chain, so $(*)$ follows immediately. Suppose $m = 2$. If $L' \subseteq L''$ or $L'' \subseteq L'$, then $(*)$ is immediate. Hence assume $L' \nsubseteq L''$ and $L'' \nsubseteq L'$. Then there exists an element $a \in L' - L' \cap L''$. Let e be the exponent of a with respect to $L_0 = L' \cap L''$. Since $a^{p^{e-1}} \notin L_0$, $a^{p^{e-1}} \notin L''$. Hence a has exponent e with respect to L''. Thus $L_0(a)$ and L'' are linearly disjoint over L_0. Hence $L'L''/L_0$ has multiplicity 1 larger than the multiplicity of L''/L_0. Thus L''/L_0 is a simple extension. Similarly L'/L_0 is a simple extension. Therefore L' and L'' are linearly disjoint over

L_0, which means that (*) holds. Thus $\mathcal{L}$ is a modular lattice. Finally, suppose $m > 2$. Let $\{a_1, \ldots, a_m\}$ be a minimal generating set of L/K and set $K' = K(L^p)$. Consider the intermediate fields $L' = K'(a_1, a_2)$ and $L'' = K'(a_1 + a_2 a_3, a_3)$. Obviously $[L'L'' : L'] = p$. Since $a_1 + a_2 a_3 \notin K'(a_3)$, $[L'' : K'] = p^2$. By using the linear independence of the power products $a_1^{i_1} a_2^{i_2} a_3^{i_3} (0 \leq i_j < p,\ j = 1,2,3)$ one can show that $L' \cap L'' = K'$, so (*) does not hold. $\mathcal{L}$ is therefore not a modular lattice.

q.e.d.

<u>2.5. Proposition</u>. Let L/K be a 2-fold purely inseparable extension. If L/K has a modular base, then L'/K has a modular base for every intermediate field L' of L/K.

Proof. L'/K is 1-fold or 2-fold. If L'/K is 1-fold, then L'/K is simple and thus has a modular base. Suppose L'/K is 2-fold. It is readily verified that for every i, $L_i (= K^{p^{-i}} \cap L)$ has a subbase. Select i such that $L_i \supseteq L'$ and L_i/K and L'/K have equal exponents. Hence $L' \not\subseteq K(L_i^p)$. So if $L' = K(a,b)$, we may assume $L_i = K(a,c)$ where a has the larger canonical exponent, namely i, and let c have the other canonical exponent, say f, so $f \leq i$. By Proposition 1.56 (4), $c^{p^f} \in (K \cap L_i^{p^f})(a^{p^f})$, so we can express c as $c = \sum \lambda_s a^s$ where $\lambda_s^{p^f} \in K$. Hence $L_i = K(a, \{\lambda_s\}) = K(a, \lambda)$, where $\lambda^{p^f} \in K$. Thus $\{a, \lambda\}$ is a canonical generating system of L_i/K with a^{p^i} and λ^{p^f} both in K, which means that $L_i = K(a) \otimes_K K(\lambda) = K(a)(\lambda) \supseteq L' = K(a)(\lambda^{p^r}) = K(a) \otimes_K K(\lambda^{p^r})$ for some $r > 0$. q.e.d.

2.6. Lemma. Let L/K and L'/K be purely inseparable extensions. Suppose $L = K(b)$ and $[L : K] = p^e$, $e > 0$. If f is an isomorphism of the lattice $\mathcal{L}$ of L/K onto the lattice $\mathcal{L}'$ of L'/K, then

(a) there exists $b' \in L'$ such that $L' = K(b')$ and $[L : K] = p^e$,

(b) $f(K(b^{p^i})) = K(b'^{p^i})$, $i = 0,1,\ldots,e$.

Proof. (a) Since L/K is simple with exponent e, there exist only $e - 1$ proper intermediate fields in L'/K. Furthermore, K is an infinite field because $e > 0$. Thus, by a well-known fact that implies a simple extension, there exists $b' \in L'$ such that $L' = K(b')$. If $e' = \exp b'/K$, then our hypothesis on f implies $e' = e$. (If $S \subseteq L$, then we denote the exponent of $K(S)/K$ by $\exp S/K$.)

(b) This follows from the fact that $\mathcal{L}$ and $\mathcal{L}'$ are finite chains of equal length. q.e.d.

2.7. Lemma. Let L/K and L'/K be purely inseparable. Suppose there exists an isomorphism f of $\mathcal{L}$ onto $\mathcal{L}'$.

(a) If L/K has a modular base B, then L'/K has a modular base B' and $|B| = |B'|$. A $1 - 1$ mapping of B

onto B' can be chosen so that if $b \leftrightarrow b'$, then $f(K(b)) = K(b')$.

(b) If L/K has a minimal generating set M, then L'/K has a minimal generating set M' and $|M| = |M'|$. A $1 - 1$ mapping of M onto M' can be chosen so that if $m \leftrightarrow m'$, then $f(K(m)) = K(m')$.

Proof. (a) By Lemma 2.6, if $b \in B$ there exists $b' \in L'$ such that $f(K(b)) = K(b')$ and $\exp b/K = \exp b'/K$. As b ranges over B it determines a subset B' of L' such that to every $b' \in B'$ there corresponds $b \in B$ for which $f(K(b)) = K(b')$. Since a modular base is a minimal generating set and f is $1 - 1$, the correspondence $b \to b'$ is $1 - 1$, hence $|B| = |B'|$. Also, because $f(L) = L'$, $L = \text{lub}\{K(b) \mid b \in B\}$ implies $L' = \text{lub}\{K(b') \mid b' \in B'\} = K(B')$. Now, if $b_0 \leftrightarrow b_0'$, then $K(B - b_0) = \text{lub}\{K(b) \mid b \in B - b_0\}$ implies $K(B' - b_0') = \text{lub}\{K(b') \mid b' \in B' - b_0'\}$. Since $L' = K(B' - b_0')(b_0')$, Lemma 2.6 implies that $\exp b_0/K(B - b_0) = \exp b_0'/K(B' - b_0')$. Hence B' is a modular base of L'/K. The proof of condition (b) is similar to that of condition (a). q.e.d.

<u>2.8. Definition</u>. We say that two purely inseparable extensions L/K and L'/K have the same canonical invariants if and only if L/K and L'/K have canonical systems $\{B_1, B_2, \ldots\}$ and $\{B_1', B_2', \ldots\}$, respectively, such that $|B_i| = |B_i'|$, $i = 1, 2, \ldots$.

2.9. Proposition. Suppose the lattices $\mathcal{L}$ of L/K and $\mathcal{L}'$ of L/K are isomorphic. If L/K has a canonical generating system, then L'/K has a canonical generating system and the two extensions have the same canonical generating invariants.

Proof. Let f be a lattice isomorphism of $\mathcal{L}$ onto $\mathcal{L}'$ and let $\{B_1, B_2, \ldots\}$ be a canonical generating system of L/K. Set $M = \bigcup_{i=1}^{\infty} B_i$ and choose $M' \subseteq L'$ as in Lemma 2.7. Then, for $i = 1,2,\ldots,$ choose $B_i' \subseteq M'$ such that $b' \in B_i'$ if and only if $f(K(b)) = K(b')$ for $b \in B_i$. Then M' is a minimal generating set of L'/K and $|B_i| = |B_i'|$, $i = 1,2,\ldots$. Set $K_i = K(B_{i+1}, B_{i+2}, \ldots) = \text{lub}\{K(b) \mid b \in B_{i+1} \cup B_{i+2} \cup \ldots\}$ and $K_i' = K(B_{i+1}', B_{i+2}', \ldots) = \text{lub}\{K(b') \mid b' \in B_{i+1}' \cup B_{i+2}' \cup \ldots\}$. Since for each $b' \in B_j'$ there exists $b \in B_j$ such that $f(K(b)) = K(b')$, we have $fK_i = K_i'$, $i = 1,2,\ldots$. By a similar argument, for any $b' \in B'$ $f(K_i(B_i - b)) = K_i'(B_i' - b')$ where $f(K(b)) = K(b')$. By Lemma 2.6, $\exp b/K_i(B_i - b) = \exp b'/K_i'(B_i' - b')$. Thus $\{B_1', B_2', \ldots\}$ is a generating system of L'/K which satisfies all the conditions of (a) in Corollary 1.31 except possibly for condition (1). To show the latter is also satisfied, note that by Lemma 2.6,

$f(K(B_{i+1}^{p^i}, B_{i+2}^{p^i}, \ldots)) = f(\text{lub}\{K(b^{p^i}) \mid b \in B_{i+1} \cup B_{i+2} \cup \ldots\} = \text{lub}\{K(b'^{p^i}) \mid b' \in B_{i+1}' \cup B_{i+2}' \cup \ldots\} = K(B_{i+1}'^{p^i}, B_{i+1}'^{p^i}, \ldots)$. Also, for all $b' \in B_i'$, $f(K(B_{i+1}^{p^i}, B_{i+2}^{p^i}, \ldots)(b)) = K(B_{i+1}'^{p^i}, B_{i+2}'^{p^i}, \ldots)(b')$

where $f(K(b)) = K(b')$. From Lemma 2.6, $b'^{p^i} \in K(B'^{p^i}_{i+1}, B'^{p^i}_{i+2}, \ldots)$. q.e.d.

<u>2.10. Lemma</u>. Let f be an isomorphism of the lattice $\mathcal{L}$ of L/K onto the lattice $\mathcal{L}'$ of L'/K, and let $F \subseteq G$ be intermediate fields of L/K such that G/F has an exponent. Then

(a) $\exp G/F = \exp fG/fF$,

(b) $f(F(G^p)) = (fF)((fG)^p)$,

(c) $fL_j = L'_j$, $j = 1,2,\ldots$, and f induces an isomorphism of the lattice of $L_{ij}/L_{i,j+1}$ onto the lattice of $L'_{ij}/L'_{i,j+1}$.

Proof. (a) Apply Lemma 2.6 to the restriction of f to the lattice of G/F. Then $G = \text{lub}\{F(a) \mid a \in G\}$ $\overset{f}{\rightarrow} \text{lub}\{(fF)(a') \mid a' \in fG\} = fG$, so $\max\{\exp a/F \mid a \in G\} = \max\{\exp a'/fF \mid a' \in fG\}$.

(b) $F(G^p) = \text{glb}\{H \mid H$ a field, $F \subseteq H \subseteq G,\ G^p \subseteq H\}$ $\overset{f}{\rightarrow} \text{glb}\{H' \mid H'$ a field, $fF \subseteq H' \subseteq fG,\ (fG)^p \subseteq H'\} = (fF)((fG)^p)$.

(c) $L_j = \text{lub}\{H \mid H$ a field, $K \subseteq H \subseteq L,\ H^{p^j} \subseteq K\}$ $\overset{f}{\rightarrow} L'_j$ and $L_{ij} = \text{glb}\{H \mid H$ a field, $L_{i-1} \subseteq H \subseteq L_j,\ L_j^{p^{i-1}} \subseteq H\}$ $\overset{f}{\rightarrow} L'_{ij}$. q.e.d.

<u>2.11. Proposition</u>. Let L/K and L'/K be purely inseparabl extensions with lower tower systems $\{T_{ij}\}$ and $\{T'_{ij}\}$, respectively. If $\mathcal{L}$ and $\mathcal{L}'$ are isomorphic, then there exists

a 1 - 1 mapping g of N onto N' such that

(a) $g(N_{ij}) = N'_{ij}$, $1 \leq i \leq j = 1,2,\ldots,$

(b) $\ell(x) = \ell(g(x))$ for every $x \in N$.

Hence L/K and L'/K have the same lower tower invariants.

Proof. Let f be an isomorphism of $\mathcal{L}$ onto $\mathcal{L}'$. We first construct a particular lower tower system $\{T'_{ij}\}$ of L'/K to associate with a given lower tower system $\{T_{ij}\}$ of L/K. Fix j and consider $T_{jj},\ldots,T_{1j}$. By Lemmas 2.7 and 2.10, we can select a relative p-base T'_{jj} of $L'_j/L'_{j,j+1}$ such that there exists a 1 - 1 mapping h_j of T_{jj} onto T'_{jj} with the property: for $t \in T_{jj}$ and $t' \in T'_{jj}$, $t' = h_j(t)$ if and only if $L_{j,j+1}(t) \xrightarrow{f} L'_{j,j+1}(t')$. Suppose we have constructed $T'_{jj},\ldots,T'_{ij}$ in L'/K such that for $k = i,\ldots,j-1$, we have

(1) $T'_{kj} \subseteq T'^{p}_{k+1,j}$ and T'_{kj} is a relative p-base of $L'_{kj}/L'_{k,j+1}$ and

(2) there exists a 1 - 1 mapping h_k of T_{kj} onto T'_{kj} with the property: for $t \in T_{kj}$ and $t' \in T'_{kj}$, $t' = h_k(t)$ if and only if $L_{k,j+1}(t) \xrightarrow{f} L'_{k,j+1}(t')$ with $t'^{p^{-1}} = h_{k+1}(t^{p^{-1}}) \in T'_{k+1,j}$ and $t^{p^{-1}} \in T_{k+1,j}$. In particular, we are thus assuming $|N_{k+1,j} - N_{kj}| = |N'_{k+1,j} - N'_{kj}|$ for $k = i,\ldots,j-1$.

Consider $L_{i,j+1}(t)/L_{i-1,j+1}$ where $t \in T_{ij}$. We have by the induction hypothesis that $L_{i,j+1}(t) \xrightarrow{f} L'_{i,j+1}(t')$ where $t' = h_i(t)$, and by Lemma 2.10, $L_{i-1,j+1} \xrightarrow{f} L'_{i-1,j+1}$. By Lemma 2.10, using $L^p_{i,j+1} \subseteq L_{i-1,j+1}$, we find that $L_{i-1,j+1}(t^p) =$ $L_{i-1,j+1}(L_{i,j+1}(t)^p) \xrightarrow{f} L'_{i-1,j+1}(L'_{i,j+1}(t')^p) = L'_{i-1,j+1}(t'^p)$ for all $t \in T_{ij}$ and $t' = h_i(t) \in T'_{ij}$. Since $L_{i-1,j+1}(T_{i-1,j}) =$ $\text{lub}\{L_{i-1,j+1}(t^p) \mid t^p \in T_{i-1,j}$ and $t \in T_{ij}\}$, we define $T'_{i-1,j}$ to be $\{t'^p \mid t^p \in T_{i-1,j}$ and $L_{i-1,j+1}(t^p) \xrightarrow{f} L'_{i-1,j+1}(t'^p)\}$. Hence $L_{i-1,j} \xrightarrow{f} \text{lub}\{L'_{i-1,j+1}(t'^p) \mid t'^p \in T'_{i-1,j}\} = L'_{i-1,j}$. Since f is $1-1$ and $T_{i-1,j}$ is a relative p-base, we have that $T'_{i-1,j}$ is a relative p-base. By the construction, there exists a mapping h_{i-1} satisfying (2) above with k replaced by $i-1$. That is, $|N_{ij} - N_{i-1,j}| = |N'_{ij} - N'_{i-1,j}|$.

This construction terminates with T'_{1j}. Then we define a $1-1$ mapping g_j from T_{jj} onto T'_{jj} with the desired properties with respect to $T_{jj},\ldots,T_{1j}$. Since the process is independent of the choice of j, the union of the g_j's defines a mapping of N onto N' with the desired properties, except that $\{T'_{ij}\}$ was not an arbitrary lower tower system. By Proposition 1.42, there exists a composite mapping that has the desired properties when $\{T'_{ij}\}$ is arbitrary. q.e.d.

<u>2.12. Proposition</u>. Suppose L/K and L'/K are purely eparable extensions and $\mathcal{L} \cong \mathcal{L}'$.

(a) If L/K has a lower tower generating system (set), n L'/K has a lower tower generating system (set) with the e lower tower invariants.

(b) L/K and L'/K have the same canonical invariants and same upper tower invariants. (See Proposition 1.50.)

Proof. (a) By Lemma 2.7, T_j' is a minimal generating of L_j'/L_{j-1}' because T_j is one of L_j/L_{j-1}.

(b) The conclusion follows here by use of Lemma 2.10 and uments similar to those used in Proposition 2.9. q.e.d.

<u>2.13. Example</u>. When L/K and L'/K have the same canonl generating invariants, their respective lattices $\mathcal{L}$ and need not be isomorphic: Let $K = P(x,y,z)$, $L = K(z^{p^{-2}}, y^{p^{-1}})$ $L' = K(z^{p^{-2}}, z^{p^{-2}}y^{p^{-1}} + x^{p^{-1}})$ where P is a perfect field x,y,z are independent indeterminates over P. Then $^{-2}, y^{p^{-1}}\}$ is a modular basis of L/K, but L'/K does not e a modular basis. Hence by Lemma 2.7, the lattices cannot isomorphic. Now, $\{z^{p^{-2}}, y^{p^{-1}}\}$ and $\{z^{p^{-2}}, z^{p^{-2}}y^{p^{-1}} + x^{p^{-1}}\}$ respective canonical generating systems with the same canonl invariants. That L'/K does not have a modular base we know Example 1.59, but it also follows from (4) of Proposition 1.51, for $^{-2}y^{p^{-1}} + x^{p^{-1}})^p \notin (L'^p \cap K)(z^{p^{-2}})^p)$. ($L/K$ and L'/K also

have the same upper tower invariants.)

2.14. Example. When L/K and L'/K have the same lower tower generating invariants, their respective lattices $\mathfrak{L}$ and $\mathfrak{L}'$ need not be isomorphic: Let $K = P(x,y,z,w)$, $L = K(z^{p^{-3}}, z^{p^{-3}}x^{p^{-1}} + y^{p^{-1}}, w^{p^{-1}})$ and $L' = K(z^{p^{-3}}, z^{p^{-3}}x^{p^{-1}} + y^{p^{-2}}, y^{p^{-1}})$ where P is a perfect field and x,y,z,w are independent indeterminates over P. Since $y^{p^{-1}}$ is not needed to generate L'/K, it follows that L/K has a minimal generating set with 3 elements while L'/K has one with 2 elements. Hence, by Lemma 2.7, the lattices cannot be isomorphic. Now $\{z^{p^{-3}}, z^{p^{-3}}x^{p^{-1}} + y^{p^{-1}}, w^{p^{-1}}\}$ and $\{z^{p^{-3}}, z^{p^{-3}}x^{p^{-1}} + y^{p^{-2}}, y^{p^{-1}}\}$ are respective lower tower generating sets and $T_{33} = \{z^{p^{-3}}, z^{p^{-3}}x^{p^{-1}} + y^{p^{-1}}\}$, $T_{23} = \{z^{p^{-2}}\}$, $T_{13} = \{z^{p^{-1}}\}$, $T_{22} = T_{12} = \emptyset$, $T_{11} = \{w^{p^{-1}}\}$ and $T'_{33} = \{z^{p^{-3}}, z^{p^{-3}}x^{p^{-1}} + y^{p^{-2}}\}$, $T'_{23} = \{z^{p^{-2}}\}$, $T'_{13} = \{z^{p^{-1}}\}$, $T'_{22} = T'_{12} = \emptyset$, $T'_{11} = \{y^{p^{-1}}\}$. Thus L/K and L'/K have the same lower tower invariants. In fact, there exists a mapping g of the type described in Proposition 2.11. q.e.d.

2.15. Example. When L/K and L'/K have the same upper and lower tower invariants, their respective lattices $\mathfrak{L}$ and $\mathfrak{L}'$ need not be isomorphic: Let $K = P(x,y,z,w)$, $L = K(z^{p^{-4}}, z^{p^{-4}}x^{p^{-1}} + y^{p^{-3}}, w^{p^{-1}})$ and $L' = K(z^{p^{-4}}, z^{p^{-4}}x^{p^{-2}} + y^{p^{-3}}, x^{p^{-1}})$ where P is a perfect field and x,y,z,w are independent indeterminates over P. It is easy to verify that L/K and L'/K

have the same lower tower invariants and the same upper tower invariants from the following lists. L/K and L'/K have the following lower tower systems $\{T_{ij}\}$ and $\{T'_{ij}\}$, respectively: $T_{44} = \{z^{p^{-4}}, z^{p^{-4}}x^{p^{-1}} + y^{p^{-3}}\}$, $T_{34} = \{z^{p^{-3}}\}$, $T_{24} = \{z^{p^{-2}}\}$, $T_{14} = \{z^{p^{-1}}\}$, $T_{33} = T_{23} = T_{13} = \emptyset$, $T_{22} = \{y^{p^{-2}}\}$, $T_{12} = \{y^{p^{-1}}\}$, $T_{11} = \{w^{p^{-1}}\}$ and $T'_{44} = \{z^{p^{-4}}, z^{p^{-4}}x^{p^{-2}} + y^{p^{-3}}\}$, $T'_{34} = \{z^{p^{-3}}\}$, $T'_{24} = \{z^{p^{-2}}\}$, $T'_{14} = \{z^{p^{-1}}\}$, $T'_{33} = T'_{23} = T'_{13} = \emptyset$, $T'_{22} = \{y^{p^{-2}}\}$, $T'_{12} = \{y^{p^{-1}}\}$, $T'_{11} = \{x^{p^{-1}}\}$.

L/K and L'/K have the following upper tower systems $\{M_{ij}\}$ and $\{M'_{ij}\}$, respectively: $M_{14} = \{z^{p^{-4}}, z^{p^{-4}}x^{p^{-1}} + y^{p^{-3}}\}$, $M_{13} = M_{12} = \emptyset$, $M_{11} = \{w^{p^{-1}}\}$, $M_{24} = \{z^{p^{-3}}, z^{p^{-3}}x + y^{p^{-2}}\}$, $M_{23} = M_{22} = \emptyset$, $M_{34} = \{z^{p^{-2}}, z^{p^{-2}}x^{p} + y^{p^{-1}}\}$, $M_{33} = \emptyset$, $M_{44} = \{z^{p^{-1}}\}$ and $M'_{14} = \{z^{p^{-4}}, z^{p^{-4}}x^{p^{-2}} + y^{p^{-3}}\}$, $M'_{13} = M'_{12} = \emptyset$, $M'_{11} = \{x^{p^{-1}}\}$, $M'_{24} = \{z^{p^{-3}}, z^{p^{-3}}x^{p^{-1}} + y^{p^{-2}}\}$, $M'_{23} = M'_{22} = \emptyset$, $M'_{34} = \{z^{p^{-2}}, z^{p^{-2}}x + y^{p^{-1}}\}$, $M'_{33} = \emptyset$, $M'_{44} = \{z^{p^{-1}}\}$.

Now $K(L^p) = K(z^{p^{-3}}, z^{p^{-3}}x + y^{p^{-2}}) = K(z^{p^{-3}}, y^{p^{-2}})$ and $K(L'^p) = K(z^{p^{-3}}, z^{p^{-3}}x^{p^{-1}} + y^{p^{-2}})$. Hence $K(L^p)/K$ has a modular base while $K(L'^p)/K$ does not. Thus by Lemma 2.7 the intermediate field lattices of $K(L^p)/K$ and $K(L'^p)/K$ are not isomorphic. Hence $\mathcal{L}$ and $\mathcal{L}'$ are not isomorphic by Lemma 2.10.

B. <u>More on type R</u>. We now extend the investigation started in I B.

<u>2.16. Definition</u>. Let $\mathcal{L}$ be the intermediate field lattice of L/K. Set $S(L/K) = \{L' \mid L' \in \mathcal{L}\}$. Let $C(L/K)$ denote a subset of $S(L/K)$. L/K is said to be of type $R(C)$ if and only if for all $L' \in C(L/K)$, L'/K is of type R.

When K is the fixed ground field, we often write $S(L)$ and $C(L)$ for $S(L/K)$ and $C(L/K)$, respectively.

<u>2.17. Definition</u>. Let $\mathcal{L}$ be the intermediate field lattice of L/K. In addition to the already defined sets

$$S(L) = \{L' \mid L' \in \mathcal{L}\},$$

$$D(L) = \{L' \mid L' \in S(L),\ L' \text{ is distinguished in } L/K\},$$

$$F_c(L) = \{L' \mid L' \in S(L),\ [L : L'] < \infty\}.$$

we also define

$$F(L) = \{L' \mid L' \in S(L),\ [L' : K] < \infty\},$$

$$B_c(L) = \{L' \mid L' \in S(L),\ L/L' \text{ has an exponent}\},$$

$$B(L) = \{L' \mid L' \in S(L),\ L'/K \text{ has an exponent}\},$$

$$N_c(L) = \{L' \mid L' \in S(L),\ L/L' \text{ is modular}\},$$

$$N(L) = \{L' \mid L' \in S(L),\ L'/K \text{ is modular}\},$$

$$U(L) = \{L' \mid L' \in S(L),\ L'/K \text{ has no exponent}\},$$

$U_c(L) = \{L' \mid L' \in S(L),\ L/L'$ has no exponent$\}$,

$U_u(L) = \{L' \mid L' \in S(L),\ L/L'$ and L'/K have no exponent$\}$,

$G(L) = \{K^{p^{-i}} \cap L \mid i = 0,1,\ldots\}$,

$E(L) = \{K(L^{p^i}) \mid i = 0,1,\ldots\}$,

where the subscript c suggests cofinite, cobounded, comodular and counbounded.

Since $U(L) \cup B(L) = S(L)$ and L/K is always of type $R(B)$, we have by Corollary 1.18 that L/K is of type $R(U)$ if and only if L/K has an exponent. Extensions of type $R(U_c)$ and $R(U_u)$ have an exponent if and only if relatively perfect intermediate fields other than K are absent, as shown by the Corollary to the following proposition.

2.18. Proposition. Let L/K be purely inseparable. Then the following conditions are equivalent:

(1) L/K is of type $R(U_c)$.

(2) L/K is of type $R(U_u)$.

(3) For every intermediate field L' of L/K, if L/L' has no exponent then L'/K has an exponent (that is $L' \in U_c(L)$ implies $L' \in B(L)$).

Proof. (1) implies (2) because $U_u(L) \subseteq U_c(L)$. For (2) implies (3), let L' be an intermediate field of L/K such that L/L' has no exponent. Let L'' be an intermediate field of L'/K. If L''/K has no exponent, then $L'' \in U_u(L)$ so L''/K is of type R. If L''/K has an exponent, then L''/K is of type R. Hence L'/K is of type $R(S)$. Thus L'/K has an exponent by Corollary 1.18. (Hence $U_u(L) = \emptyset$.) (3) implies (1) because by (3) $U_c(L) \subseteq B(L)$ and L/K is always of type $R(B)$. q.e.d.

2.19. Corollary. Let L/K be purely inseparable. If K is the only relatively perfect intermediate field of L/K, then (1), (2) and (3) of Proposition 2.18 are each equivalent to L/K has an exponent.

Proof. Suppose condition (3) holds and L/K has no exponent. Then there exists an intermediate field L' (possibly L) of L/K such that L/L' has an exponent and L'/K is not of type R. Then there exists a relative p-base M of L'/K such that $L' \supset K(M)$. Let $L'' = K(M)$. Now L'/L'' must have no exponent, otherwise $L' = L''$. Hence L''/K has an exponent, say e, by condition (3). Thus $K(L'^{p^e}) = K(L'^{p^{e+1}})$, which contradicts the fact that K is the only relatively perfect intermediate field of L/K. Hence L/K has an exponent. q.e.d.

If L/K is of type R, then L/L' is of type R for every $L' \in S(L)$ by Proposition 1.15. Thus if L/K is of type R and K^* is an intermediate field of L/K such that L/K^* is modular, then by Proposition 1.23 L/K^* has an exponent; hence L/K^* has a modular basis by Proposition 1.56. The existence of a minimal K^* such that L/K^* is modular is proved in Proposition 1.58.

2.20. Definition. Consider K as a fixed ground field. Suppose $C(L)$ is defined for every field L containing K, $C(L) \subseteq S(L)$. We say that $C(L)$ is cofinite if and only if for all L such that L/K is of type $R(C)$ and for all $L^* \in S(L)$ such that L/L^* is finite, L^*/K is of type $R(C)$.

If $C(L) = \{L\}$, then $C(L)$ is cofinite by Proposition 1.20. In fact, $F_c(L)$ is cofinite.

2.21. Definition. Suppose L/K is of type $R(C)$. Set $L(C) = \{L' \mid L' \in S(L)$ and L'/K is of type $R(C)\}$. Then (1) L/K is of type $R^m(C)$ if and only if $L(C)$ constitutes a complete meet-semilattice in $S(L)$, and (2) L/K is of type $R^j(C)$ if and only if $L(C)$ constitutes a complete join-semilattice in $S(L)$.

2.22. Proposition. Let $L \in C(L)$. (a) Suppose $C(L)$ is cofinite. L/K is of type $R^m(C)$ if and only if L/K is of type $R(S)$. (b) Suppose that for $L' \in S(L)$, L'/K is finitely generated implies L'/K is of type R. L/K is type $R^j(C)$ if and only if L/K is of type $R(S)$.

Proof. (a) Suppose L/K is of type $R^m(C)$. Let $L' \in S(L)$. Since $L \in C(L)$, L/K is of type R whence L/L' is of type R by Proposition 1.15. Let M be a relative p-base of L/L' and set $L_1 = \bigcap_{m \in M} L'(M - m)$. Since $L'(M - m)$ is cofinite in L/K, L_1 is the intersection of fields of type $R(C)$ (over K). Thus L_1/K is of type $R(C)$. If

$L_1 \supset L'$, then since L_1/K is of type R we can continue this process, replacing L by L_1. Hence there exists a well-ordered set of intermediate fields of L/L', namely $\{L, L_1, \ldots\}$, each of type $R(C)$ such that their intersection is L'. Therefore L'/K is of type $R(C)$ whence of type R. Hence L/K is of type $R(S)$. (If L/K is purely inseparable, then L/K is of type $R^j(C)$ if and only if L/K has an exponent by Corollary 1.18.) The converse is obvious.

(b) We first show that L/K is of type $R^j(C)$ if and only if for all $L' \in S(L)$, L'/K is of type $R(C)$. Suppose L/K is of type $R^j(C)$. If $L' \in S(L)$, then $L' = \text{lub}\{K(a) \mid a \in L'\}$. Since each $K(a)/K$ is of type $R(C)$, L'/K is of type $R(C)$. Conversely, suppose for all $L' \in S(L)$, L'/K is of type $R(C)$. Then, in particular, L'/K is of type $R(C)$ for those L' which are composites of fields of type $R(C)$. Hence we see that if $L \in C(L)$, then L/K is of type $R^j(C)$ if and only if L/K is of type $R(S)$. (If L/K is purely inseparable, then L/K is of type $R^m(C)$ if and only if L/K has an exponent.) q.e.d.

2.23. Corollary. Suppose L/K is purely inseparable. The following types are identical, and L/K is of this type if and only if L/K has an exponent: $R^m(B_c)$, $R^m(F_c)$ $R^m(E)$, $R^j(B_c)$, $R^j(F_c)$, $R^j(E)$ and $R^j(N_c)$.

Proof. Each of the sets $B_c(L)$, $F_c(L)$, $E(L)$ are cofinite and each contains L. q.e.d.

2.24. Corollary. Suppose L/K is purely inseparable. The three types $R(C)$, $R^m(C)$ and $R^j(C)$ are identical when $C = N, B, F$ or G.

Proof. For every subset $B'(L)$ of $B(L)$ and for every $L' \in S(L)$, L'/K is always of type $R(B')$. Thus if L' is the intersection or composite of fields of type $R(B')$, then L'/K is of type $R(B')$. Therefore the three types $R(B')$, $R^m(B')$ and $R^j(B')$ are identical. If L/K is of type $R(N)$, then $N \subseteq B$ by Proposition 1.23. q.e.d.

2.25. Proposition. Suppose L/K is purely inseparable. Then the three type $R(C)$, $R^m(C)$ and $R^j(C)$ are identical when $C = U_c$ or U_u.

Proof. Suppose L/K is of type $R(U_c)$. Let $L' \in S(L)$. If L'' is any element of $U_c(L')$, then $L'' \in U_c(L)$. Hence L''/K is of type R. Thus L'/K is of type $R(U_c)$. Since L' was arbitrary, L/K is of type $R^j(U_c)$ as in the proof of (b) of Proposition 2.22. Hence $R(U_c)$ and $R^j(U_c)$ are identical. By Proposition 2.18, the intersection of any collection of fields in $U_c(L)$ is again in $U_c(L)$, in fact of bounded exponent over K. Hence L/K is of type $R^m(U_c)$.

Suppose L/K is of type $R(U_u)$. Then $U_u(L) = \emptyset$ by Proposition 2.18 whence L/K is of type $R^m(U_u)$. Also for all $L' \in S(L)$, $U_u(L') = \emptyset$. Hence for all $L' \in S(L)$, L'/K is of type $R(U_u)$ so that in the proof of (b) of Proposition 2.22, L/K is of type $R^j(U_u)$. q.e.d.

Reference Note for Chapter II

The result 2.4 is due to Pickert [45]. The lattice invariants are treated in [15], [17]. The additional type R material is in [37]. Based on the fact that a proper purely inseparable extension has non-trivial derivations, a Galois theory has been emerging; some of the results of this theory can be found in sources included in our reference list, in particular, [5], [13], [22], [24], [55], [57].

III. Some Applications

A. Extension coefficient fields. We now consider for certain commutative algebras the connection between our generating systems and the existence of coefficient fields containing the base field.

Let A be a commutative algebra with identity over the field $K(\subseteq A)$. If N is a fixed maximal ideal of A and if g is the natural algebra epimorphism of A onto A/N, we identify K and gK in A/N and sometimes denote A by (A,K,N,g).

We give a direct definition of the type of coefficient fields with which we are concerned. The well-known definition and existence theorems in the case of complete local rings (see for example, [7] and [40]) do not in general imply the existence of our type.

3.1. Definition. (A,K,N,g) is said to have an extension coefficient field (or, K-coefficient field) F for N if and only if there exists a field F in A such that $gF = gA$ and $K \subseteq F$.

Even the tensor product $K(a) \otimes F$, where $K(a)$ is a simple purely inseparable extension of exponent $e > 1$ and F is a proper intermediate field of $K(a)/K$ does not possess a $1 \otimes F$ - coefficient field. (Here and below, $\otimes$ means $\otimes_K$.)

3.2. Lemma. Suppose H, K, L are fields such that $H \subseteq K \subseteq L$. There exists a subset M of L which is a relative p-base of both L/K and L/H if and only if $K(L^p) = H(L^p)$.

Proof. Suppose M exists. Then $K(L^p) = H(L^p)$ since $L/H(L^p)$ has exponent 1 and $H(L^p) \subseteq K(L^p) \subseteq L$. The converse is immediate. q.e.d.

3.3. Proposition. Let K, F, L be fields such that $K \subseteq F \subseteq L$ and L/F has exponent e. Then the following conditions are equivalent.

(1) There exists an intermediate field J of L/K such that $L \cong J \otimes F$, J/K is modular and L/F is modular ($\otimes = \otimes_K$ as usual).

(2) There exists a canonical generating system $\{B_1, \ldots, B_e\}$ of L/F such that $B_i^{p^i} \subseteq (L^{p^i} \cap K)(K_i^{p^i})$ where $K_i = F(B_{i+1}, \ldots, B_e)$, $i = 1, \ldots, e$.

(3) For all canonical generating systems $\{B_1, \ldots, B_e\}$ of L/F, $B_i^{p^i} \subseteq (L^{p^i} \cap K)(K_i^{p^i})$, $i = 1, \ldots, e$.

Proof. (1) implies (2): Since $L \cong J \otimes F$, J/K also has exponent e and every canonical generating system of J/K is one for L/F. Since J/K is modular, J/K has a canonical generating system $\{A_1, \ldots, A_e\}$ such that $A_1 \cup \ldots \cup A_e$ is a modular base of J/K. Clearly $\{A_1, \ldots, A_e\}$ satisfies (2).

(2) implies (3): Let $\{B_1,\ldots,B_e\}$ be a canonical generating system of L/F satisfying the conditions in (2). Set $K_i = F(B_{i+1},\ldots,B_e)$, $i = 1,\ldots,e$, where K_e means F. Then $L^p = K_1^p(B_1^p)$ so that $L^p = (L^p \cap K)(K_1^p)(B_1^p) = (L^p \cap K)(K_1^p)$. Make the induction hypothesis that $L^{p^i} = (L^{p^i} \cap K)(K_i^{p^i})$. Then $L^{p^{i+1}} = (L^{p^{i+1}} \cap K^p)(K_i^{p^{i+1}}) \subseteq (L^{p^{i+1}} \cap K)(K_{i+1}^{p^{i+1}}, B_{i+1}^{p^{i+1}}) = (L^{p^{i+1}} \cap K)(K_{i+1}^{p^{i+1}}) \subseteq L^{p^{i+1}}$. Hence by induction $L^{p^i} = (L^{p^i} \cap K)(K_i^{p^i})$, $i = 1,\ldots,e$, whence

$$L^{p^i} = (L^{p^i} \cap K)(F^{p^i})(B_{i+1}^{p^i},\ldots,B_e^{p^i}),\ i = 1,\ldots,e.$$

Thus

$$L^{p^i} = (L^{p^i} \cap F)(B_{i+1}^{p^i},\ldots,B_e^{p^i}),\ i = 1,\ldots,e.$$

Hence, by Proposition 1.56, if $\{B_1',\ldots,B_e'\}$ is any canonical generating system of L/F, then

$$L^{p^i} = (L^{p^i} \cap F)(B_{i+1}'^{p^i},\ldots,B_e'^{p^i}),\ i = 1,\ldots,e.$$

Now $B_{i+1}^{p^i} \cup \ldots \cup B_e^{p^i}$ is a relative p-base of $L^{p^i}/(L^{p^i} \cap K)(F^{p^i})$ and $L^{p^i}/(L^{p^i} \cap F)$. By Lemma 3.2, it follows that since $B_{i+1}'^{p^i} \cup \ldots \cup B_e'^{p^i}$ is a relative p-base of $L^{p^i}/(L^{p^i} \cap F)$, it is also one for $L^{p^i}/(L^{p^i} \cap K)(F^{p^i})$. Thus $L^{p^i} = (L^{p^i} \cap K)(F^{p^i})(B_{i+1}'^{p^i},\ldots,B_e'^{p^i})$ because $L^{p^i}/(L^{p^i} \cap K)(F^{p^i})$ has an exponent. Hence $B_i'^{p^i} \subseteq (L^{p^i} \cap K)(F^{p^i})(B_{i+1}'^{p^i},\ldots,B_e'^{p^i}) = (L^{p^i} \cap K)(K_i^{p^i})$, $i = 1,\ldots,e$.

(3) implies (1): Let $\{B_1,\ldots,B_e\}$ be any canonical generating system of L/F. Then $L^{p^i} = (L^{p^i} \cap K)(K_i^{p^i})$, $i = 1,\ldots,e$. Assume $L \cong K(A_1) \otimes \ldots \otimes K(A_{i-1}) \otimes F(B_i,\ldots,B_e)$ and $\{A_1,\ldots,A_{i-1},B_i,\ldots,B_e\}$ is a canonical generating system of L/F. Take a subset $C_i^{p^i}$ of $L^{p^i} \cap K$ such that $L^{p^i} \cap K = K^{p^i}(A_1^{p^i},\ldots,A_{i-1}^{p^i},C_i^{p^i})$. Then $L^{p^i} = K^{p^i}(A_1^{p^i},\ldots,A_{i-1}^{p^i},C_i^{p^i},B_{i+1}^{p^i},\ldots,B_e^{p^i})(F^{p^i})$. Hence $L = F(A_1,\ldots,A_{i-1},C_i,B_{i+1},\ldots,B_e)$. Select $A_i \subseteq C_i$ so that $A_1 \cup \ldots \cup A_{i-1} \cup A_i \cup B_{i+1} \cup \ldots \cup B_e$ is a relative p-base of L/F. By the induction hypothesis it follows for all $b \in B_i$ and for $F^* = F(A_1,\ldots,A_{i-1},B_{i+1},\ldots,B_e)$ that $[L : F^*(B_i - b)] = p^i$. Hence $\{B_i\}$ is a canonical generating system of L/F^* with exponent i. Since A_i is a minimal generating set of L/F^*, $\{A_i\}$ is a canonical generating system of L/F^* with exponent i by the invariance conditions of canonical generating systems. Thus, since $A_i^{p^i} \subseteq K$, $L \cong K(A_1) \otimes \ldots \otimes K(A_i) \otimes F(B_{i+1},\ldots,B_e)$. Therefore condition (1) holds by induction with $J = K(A_1,\ldots,A_e)$. (Also $L \otimes K_i$ has $K(A_1,\ldots,A_i) \otimes K_i$ as a $1 \otimes K_i$ - coefficient field, $i = 1,\ldots,e$.) q.e.d.

If F/K is separable algebraic, then Proposition 3.3 not only gives sufficient conditions for L/K to be the tensor product $J \otimes F$ of its maximal separable intermediate field F/K and the intermediate field of all purely inseparable elements J/K, but also criteria for J/K to be modular, and thus for L/K to be modular by (c) of Lemma 1.61.

We noted that (3) implies the condition:

(4) For all canonical generating systems $\{B_1, \ldots, B_e\}$ of L/F, $L \otimes K_i$ has $1 \otimes K_i$ - coefficient field, $i = 1, \ldots, e$.

Now (4) implies the following condition:

(5) There exists a canonical generating system $\{B_1, \ldots, B_e\}$ of L/K such that $L \otimes K_i$ has a $1 \otimes K_i$ - coefficient field, $i = 1, \ldots, e$.

Thus we have (1) $\Leftrightarrow$ (2) $\Leftrightarrow$ (3) $\Rightarrow$ (4) $\Rightarrow$ (5).

<u>3.4. Proposition</u>. If F/K or L/K is modular, then (1), (2), (3), (4) and (5) are equivalent.

Proof. Suppose F/K or L/K is modular. Then K and $(L^{p^i} \cap K)(F^{p^i})$ are linearly disjoint over $L^{p^i} \cap K$, $i = 1, \ldots, e$.

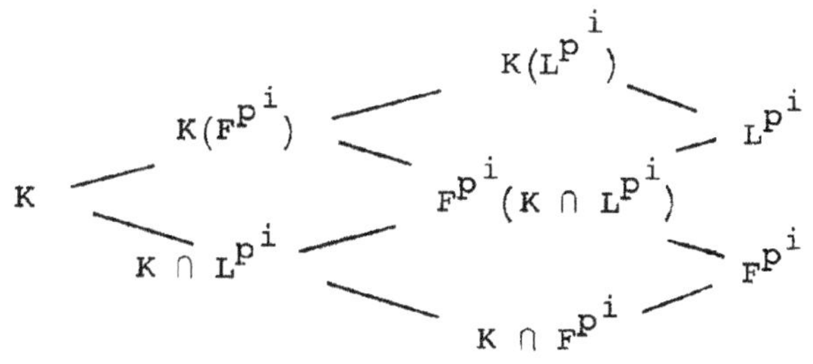

We use this property to show that (5) implies (1). Let $\{B_1, \ldots, B_e\}$ be a canonical generating system of L/K which satisfies the conditions in (5). Then $L \otimes_F K_i$ has a coefficient field containing $1 \otimes_F K_i$, $i = 1, \ldots, e$. We use the symbol $\otimes_e$ to denote the tensor product with respect to $K_{e-1} = F(B_e)$. We now show L/F has a modular base. Make the induction hypothesis that $L \otimes_F K_i$ has a $1 \otimes_F K_i$ - coefficient field $(i = 1,\ldots,e)$ implies that L/F has a modular base for all extensions L/F with exponent $< e$. There exists a canonical F-epimorphism of $L \otimes_F K_i$ onto $L \otimes_e K_i$, so $L \otimes_e K_i$ has a $1 \otimes_e K_i$ - coefficient field, $i = 1,\ldots,e-1$. Then L/K_{e-1} has a modular base, say $B_1' \cup \ldots \cup B_{e-1}'$, by the induction hypothesis. We may assume that $B_1' \cup \ldots \cup B_{e-1}'$ is canonically ordered. Since $B_1',\ldots,B_{i-1}',B_e$ satisfy the conditions in part B of Corollary 1.31, $\{B_1',\ldots,B_{e-1}',B_e\}$ is a canonical generating system of L/F. Because $L \otimes_F K_{e-1}$ has a $1 \otimes_F K_{e-1}$ - coefficient field, there exist subsets $B_j^* \subseteq L \otimes_F K_{e-1}$ such that $f\,B_j^* = B_j'$ and $B_j^{*p^j} \subseteq 1 \otimes_F K_{e-1}$ $(j = 1,\ldots,e-1)$, where f is the canonical F-epimorphism of $L \otimes_F K_{e-1}$ onto the residue field $L \otimes_F K_{e-1}$ modulo its maximal ideal. Now $B_j^{*p^j} \subseteq (L^{p^j} \otimes_F 1)[1 \otimes_F K_{e-1}^{p^j}]$, whence for all $b^* \in B_j^*$ it follows that $b^{*p^j} = \sum_s c_s^{p^j} \otimes_F b_1^{p^j s_1} \ldots b_t^{p^j s_t}$, where $c_s \in L$ and $b_1,\ldots,b_t \in B_e$. By the division algorithm, $p^j s_i = p^e\, n(s_i) + r(s_i)$, where $0 \leq r(s_i) < p^e$, $i = 1,\ldots,t$. Since $b_i^{p^e} \in F$ and p^j divides p^e, it follows that $b^{*p^j} = \sum_s c_s'^{p^j} \otimes_F b_1^{r(s_1)} \ldots b_t^{r(s_t)}$ where $c_s' \in L$. Also, with $M_j' = B_{j+1}' \cup \ldots \cup B_{e-1}' \cup B_e$, we have $fb^{*p^j} \in F(B_e) \cap F(M_j'^{p^j})$. Hence $b^{*p^j} = \sum_n k_n \otimes_F p_n$ where p_n is a monomial in the elements of $M_j'^{p^j}$ and $\{p_n\}$ is a linear basis that includes

$\{b_1^{r(s_1)} \ldots b_t^{r(s_t)}\}$. Thus each k_n equals some $c_s'^{p^j}$, whence $k_n \in L^{p^j} \cap F$. Since $fB_j^* = B_j', B_j'^{p^j} \subseteq (L^{p^i} \cap F)(M_j'^{p^j})$, $j = 1,\ldots,e-1$. Hence $\{B_1',\ldots,B_{e-1}',B_e\}$ is a canonical generating system satisfying (2) of Proposition 3.3 with $K = F$ there. Hence L/F has a modular base, say $A_1' \cup \ldots \cup A_e'$. We assume $A_1' \cup \ldots \cup A_e'$ is canonically ordered. Since $L \otimes F$ has a $1 \otimes F$ - coefficient field, there exist subsets $B_1^*,\ldots,B_e^*$ of $L \otimes F$ such that $gB_i^* = A_i'$, $i = 1,\ldots,e$, and $(1 \otimes F)[B_1^*,\ldots,B_e^*]$ is a field, where g is the canonical K-epimorphism of $L \otimes F$ onto $LF = L$. Let $b \in B_i^*$. Then $b^{p^i} \in 1 \otimes K(F^{p^i})$. By hypothesis there exists a subset $\{x_t^{p^i}\}$ of F^{p^i} which is a linear basis of both $(L^{p^i} \cap K)(F^{p^i})/(L^{p^i} \cap K)$ and $K(F^{p^i})/K$. Now $b^{p^i} = \Sigma_t k_t \otimes x_t^{p^i}, k_t \in K$. Also, $b = \Sigma_s a_s \otimes m_s$, where $a_s \in L$ and $m_s \in F$. Hence $b^{p^i} = \Sigma_s a_s^{p^i} \otimes m_s^{p^i} = \Sigma_s a_s^{p^i} \otimes (\Sigma_t k_{st} x_t^{p^i}) = \Sigma_t (\Sigma_s a_s^{p^i} k_{st}) \otimes x_t^{p^i}$, where $k_{st} \in L^{p^i} \cap K$ and $m_s = \Sigma_t k_{st} x_t^{p^i}$. Hence for each t, $k_t = \Sigma_s a_s^{p^i} k_{st} \in L^{p^i}$. Thus $A_i'^{p^i} \subseteq (L^{p^i} \cap K)(F^{p^i})$. Therefore $\{A_1',\ldots,A_e'\}$ is a canonical generating system of L/F satisfying (2) of Proposition 3.3. q.e.d.

<u>3.5. Lemma</u>. Suppose $\bigcap_{i=1}^{\infty} K(L^{p^i}) = K$. If L^{p^i} and $K(L^{p^r})$ are linearly disjoint over $L^{p^i} \cap K(L^{p^r})$, $i \leq r = 1,2,\ldots$, then L/K is modular.

Proof. Let i be fixed. Then our hypothesis implies that L^{p^i} and $K(L^{p^r})$ are linearly disjoint over their intersection for $r = i, i+1, \ldots$. Let X be a maximal subset of K which is linearly independent over $L^{p^i} \cap K(L^{p^i}) = L^{p^i}$. Then X is certainly linearly independent over each of the smaller fields $L^{p^i} \cap K(L^{p^r})$, $r = i, i+1, \ldots$. Moreover, because L^{p^i} and $K(L^{p^r})$ are linearly disjoint and $K \subseteq K(L^{p^r})$, X remains maximally linearly independent over $L^{p^i} \cap K(L^{p^r})$, $r = i, i+1, \ldots$. Also, X is clearly linearly independent over $L^{p^i} \cap K$. Finally, we show that X is a linear basis of $K/L^{p^i} \cap K$. Since $L^{p^r} \subseteq L^{p^i} \cap K(L^{p^r})$, X is a linear basis of $K(L^{p^r})/L^{p^i} \cap K(L^{p^r})$, $r = i, i+1, \ldots$. For, if not, K would not be in the linear span of X over $L^{p^i} \cap K(L^{p^r})$ and there would exist a subset $X \cup \{x\}$ in K that is linearly independent over $L^{p^i} \cap K(L^{p^r})$, which contradicts the maximality of X. Thus for any $k \in K$, $k = \sum k_{r_0} x_{r_0} = \sum k_{r_1} x_{r_1} = \ldots$ where $k_{r_j} \in L^{p^i} \cap K(L^{p^{i+j}})$ and $x_{r_j} \in X$, $j = 0, 1, \ldots$. Hence the coefficients in these linear combinations are in $\bigcap_{r=1}^{\infty} (L^{p^i} \cap K(L^{p^r})) = L^{p^i} \cap (\bigcap_{r=1}^{\infty} K(L^{p^r})) = L^{p^i} \cap K$. The desired conclusion now holds because X is linearly independent over L^{p^i}. q.e.d.

3.6. Corollary. Let L/K be purely inseparable. Suppose L/K has a canonical generating system and $\bigcap_{i=1}^{\infty} K(L^{p^i}) = K$. Then the following conditions are equivalent.

(1) There exist subsets $A_1, A_2, \ldots$ of L such that for every canonical generating system $\{B_1, B_2, \ldots\}$ of L/K and every positive integer e, $L \cong K(A_1) \otimes \ldots \otimes K(A_e) \otimes K(M_e)$ and $\{A_1, \ldots, A_e, B_{e+1}, B_{e+2}, \ldots\}$ is a canonical generating system of L/K, where $M_e = B_{e+1} \cup B_{e+2} \cup \ldots$.

(2) There exists a canonical generating system $\{B_1, B_2, \ldots\}$ of L/K such that $B_i^{p^i} \subseteq (L^{p^i} \cap K)(B_{i+1}^{p^i}, B_{i+2}^{p^i}, \ldots)$, $i = 1,2,\ldots$.

(3) For all canonical generating systems $\{B_1, B_2, \ldots\}$ of L/K, $B_i^{p^i} \subseteq (L^{p^i} \cap K)(B_{i+1}^{p^i}, B_{i+2}^{p^i}, \ldots)$, $i = 1,2,\ldots$.

(4) For all canonical generating systems $\{B_1, B_2, \ldots\}$ of L/K, $L \otimes K_i$ has a $1 \otimes K_i$ - coefficient field, where $K_i = K(M_i)$, $i = 1,2,\ldots$.

(5) There exists a canonical generating system $\{B_1, B_2, \ldots\}$ of L/K such that $L \otimes K_i$ has a $1 \otimes K_i$ - coefficient field, $i = 1,2,\ldots$.

Proof. Let e be any fixed positive integer. Then for any canonical generating system $\{B_1, B_2, \ldots\}$ of L/K, $\{B_1, \ldots, B_e\}$ is a canonical generating system of L/K_e. Set $F = K_e$. Since $(L^{p^i} \cap K)(F^{p^i})(B_{i+1}^{p^i}, \ldots, B_e^{p^i}) = (L^{p^i} \cap K)(B_{i+1}^{p^i}, B_{i+2}^{p^i}, \ldots)$, we have the equivalence of (1), (2) and (3) by Proposition 3.3. Clearly (1) implies (4) which implies (5). Now (5) implies a similar condition (5') in which $K(L^{p^e})$ replaces K and the tensor product is with respect to $K(L^{p^e})$. Letting $F = K(L^{p^e})$ in Proposition 3.4, we have that $L/K(L^{p^e})$ has a modular base. That is, (5') implies $L/K(L^{p^e})$ is modular for any positive integer e. Hence by Lemma 3.5, L/K is modular. Setting $F = K_e$, we proceed as in the proof of (3) implies (1) of Proposition 3.3 where we use the fact L/K modular implies that $L^{p^i} = (L^{p^i} \cap K)(K_i^{p^i})$, $i = 1,\ldots,e$. Hence we obtain (5) implies (1).

q.e.d.

<u>3.7. Lemma</u>. Let E/K be a field extension, F and S intermediate fields of E/K, and S/K separable algebraic. Suppose $b \in E$ is a root of the irreducible polynomial $(x^{p^e})^n + k_{n-1}(x^{p^e})^{n-1} + \ldots + k_o$ in an indeterminate x over K, $k_i \in K$, $i = 0,\ldots,n-1$, where n is the degree of separability of b over K. Suppose S, F are linearly disjoint over K and $b^{p^e} \in S$. Then $b \in SF$ if and only if $k_i^{p^{-e}} \in F$, $i = 0,\ldots,n-1$.

Proof. Suppose $b \in SF$. Since $b^{p^e} \in S$, $K(b^{p^e})$ and F are linearly disjoint over K. Thus $[F(b^{p^e}) : F] = [K(b^{p^e}) : K] = n$. Hence $[F(b) : F] = [F(b) : F(b^{p^e})] \cdot n$. Since S/K is separable algebraic, SF/F is separable algebraic.

nce $b \in F(b^{p^e})$ and $[F(b) : F] = n$. Therefore b is a ot of an irreducible polynomial of the form

$+ a_{n-1}x^{n-1} + \ldots + a_0$ over F where $a_i \in F$, $i = 0,\ldots,n-1$.

t F^* denote the composite $FK^{p^{-e}}$ in $E(K^{p^{-e}})$. Then S are linearly disjoint over K. Similarly as in the above gument, $[F^*(b) : F^*] = n$. Thus it follows that b is a root the irreducible polynomial $q(x) = x^n + k_{n-1}^{p^{-e}}x^{n-1} + \ldots + k_0^{p^{-e}}$ er $F^* \supseteq F$. Hence $k_i^{p^{-e}} = a_i \in F$, $i = 0,\ldots,n-1$.

Conversely, suppose $k_i^{p^{-e}} \in F$. Then b is a root of $q(x)$ er F. Since $x^n + k_{n-1}x^{n-1} + \ldots + k_0$ is irreducible and parable over K and $K \cong K^{p^{-e}}$, $q(x)$ is irreducible and parable over $K(k_0^{p^{-e}},\ldots,k_{n-1}^{p^{-e}})$. Hence b is separable gebraic over F. Since b is also purely inseparable over $b \in SF$. q.e.d.

If E/K is finite, F the intermediate field of all elements rely inseparable over K and S the maximal separable termediate field, then Proposition 3.1 in [47, p.221] shows at $E \not\cong S \otimes F$ implies that E/F always gives rise to ceptional extensions. Lemma 3.7 gives a necessary and fficient for $E \cong S \otimes F$.

3.8. Proposition. Let L/K be algebraic, S maximal separable termediate field of L/K, and F/K a field extension. Let G note a set of generators of L/S and for $b \in G$ let

$(x^{p^e})^n + k_{n-1}(x^{p^e})^{n-1} + \ldots + k_o$ denote the monic irreducible polynomial over K which has b as a root ($k_i \in K$ and k_i, e, n functions of b, $i = 1,\ldots,n-1$). Then (1) LF is a unique composite and (2) for all $b \in G$, $k_i^{p^{-e}} \in F$, $i = 1,\ldots,n-1$ if and only if (3) $S \otimes F$ is a coefficient field of $L \otimes F$.

Proof. Suppose (1) and (2) hold. By (1) $S \otimes F$ is a field. For any $b \in G$, $b \in SF$ by (2) and Lemma 3.7. Hence $LF = SF$. Conversely suppose (3) holds. Then S and F in LF are linearly disjoint over K. Hence (2) holds by Lemma 3.7. Since L/S is purely inseparable and $S \otimes F$ is a field, we have that (1) holds. q.e.d.

<u>3.9. Corollary</u>. Suppose LF is a unique composite. If (1) $L^{p^e} \subseteq S$ and $F^{p^e} \supseteq K$, or (2) $F^{p^\infty} \supseteq K$, then $S \otimes F$ is a coefficient field of $L \otimes F$.

<u>3.10. Proposition</u>. Let L, K, S, F and G be defined as in Proposition 3.8. Let J be an intermediate field L/K and let f denote the canonical K-epimorphism of $L \otimes F$ onto a field composite LF. Suppose $L \cap F \supseteq S$ in LF and $f(S \otimes 1) = f(1 \otimes S)$. Then (1) $L \otimes F$ has no nilpotent elements, (2) JF (in LF) is a unique composite and (3) for all $b \in G$, $k_i^{p^{-e}} \in J$, $i = 1,\ldots,n-1$, if and only if (4) $J \otimes F$ is a coefficient field of $L \otimes F$ for $\operatorname{Ker} f$ and (5) $L = JS$.

Proof. Suppose (1), (2) and (3) hold. Then (1) and (2) imply $J \otimes F$ is a field. Thus J and F (in LF) are linearly disjoint over K. Hence J and S (in L) are linearly disjoint over K. By Lemma 3.7, (3) implies (5). Thus $LF = JF$ since $f(S \otimes 1) = f(1 \otimes S)$. Hence (4) holds. Conversely, (4) implies that J and F are linearly disjoint over K. Hence (5) implies (3) by Lemma 3.7. Thus it follows that L/J is separable algebraic. Since $J \otimes F$ is a field, $L \otimes_J (J \otimes F) \cong L \otimes F$ has no nilpotent elements. That is, (1) holds. q.e.d.

<u>3.11. Corollary</u>. Suppose $L \otimes F$ has no nilpotent elements and JF is a unique composite. If (1) $L^{p^e} \subseteq S$ and $J^{p^e} \supseteq K$, or (2) $J^{p^\infty} \supseteq K$, then $J \otimes F$ is a coefficient field of $L \otimes F$ for $\operatorname{Ker} f$.

If L/K is finite and normal, then L satisfies (5) of Proposition 3.10 where J/K is purely inseparable. Thus (2) and (3) of Proposition 3.10 hold. Hence (1) and (4) are equivalent in this case.

If J is the intermediate field of L/K consisting of all elements of L which are purely inseparable over K, then (5) of Proposition 3.10 holds when L/J is not exceptional (Proposition 3.1 in [47, p.221]).

<u>3.12. Proposition</u>. Let (A,K,N,g) be a complete local algebra (not necessarily Noetherian). If A/N is modular over K, then A has an extension coefficient field if and only if $g(A^{p^i} \cap K) = (A/N)^{p^i} \cap K$, $i = 1,2,\ldots$.

Proof. Suppose that $g(A^{p^i} \cap K) = (A/N)^{p^i} \cap K$, $i = 1,2,\ldots$. Since A/N is modular over K, there exists a p-base B of A/N such that $K = ((A/N)^{p^i} \cap K)(C^*)$, $i = 1,2,\ldots$, $C^* = C \cap K$ and $C = \{b^{p^i} \mid b \in B,\ i$ is the exponent of b over K if b is purely inseparable over and $i = 0$ otherwise$\}$. (See (*) in the proof of (2) implies (3) in Proposition 1.23.) Since $g(A^{p^i} \cap K) = (A/N)^{p^i} \cap K$, there exists a set of representatives B' in A of B such that $A^{p^i}[B']$ contains C^* (K and gK being identified). Since $g(A^{p^i} \cap K) = (A/N)^{p^i} \cap K$, we have $K = (A^{p^i} \cap K)(C^*)$, $i = 1,2,\ldots$. Thus $A^{p^i}[B'] \supseteq K$, $i = 1,2,\ldots$, whence with respect to the N-adic topology of A, $\bigcap_{i=1}^{\infty}$ (closure $A^{p^i}[B']$) is a coefficient field of A containing K, [58, p.306]. The converse is immediate. q.e.d.

When A/N has no purely inseparable elements over K, then the condition $g(A^{p^i} \cap K) = (A/N)^{p^i} \cap K$ always holds since $(A/N)^{p^i} \cap K = K^{p^i}$ in this case. Also, if A/N is separable over K, then A/N is modular over K.

The existence of an extension coefficient field is a particular case of the extension of a coefficient field of a subring to the whole ring [40]. We now give a few results of this point of view.

3.13. Proposition. Let A be a commutative ring with identity, N a maximal ideal of A and g the natural homomorphism of A onto A/N. Let R be a complete local ring (not necessarily Noetherian) of prime characteristic p such that $R \subseteq A$, the identities of A and R coincide and $M = R \cap N$ is the unique maximal ideal of R. Suppose A/N is purely inseparable and has a modular base over R/M. Then there exists a coefficient field of R which is extendable to one of A if and only if $g(A^{p^i} \cap R) = (A/N)^{p^i} \cap R/M$, $i = 1,2,\ldots$.

Proof. By Proposition 1.22, there exists a p-base B of A/N such that C is a p-base of R/M. Suppose $g(A^{p^i} \cap R) = (A/N)^{p^i} \cap R/M$, $i = 1,2,\ldots$. Then there exists a set of representatives B' in A of B such that $R \supseteq C'$ where $C' = \{b'^{p^i} \mid b' \in B,\ i$ is the exponent of $b = gb'$ over $R/M\}$. Since C is a p-base of R/M, R has a coefficient field $K \supseteq C'$ by the existence lemma as stated in [39]. Since $B - (R/M)$ is a modular base of A/N over R/M and $b' \in B'$ has the same exponent over K that gb' has over R/M, we see that $K[B']$ is a coefficient field of A. The converse is immediate. q.e.d.

3.14. Corollary. Let A, N, R, M and g be defined as in Proposition 3.13. If $R \subseteq A^{p^e}$ and $gR = (A/N)^{p^e}$ for some positive integer e, then every coefficient field of R is extendable to one of A.

Proof. Let K be a coefficient field of R. Then $gK = (A/N)^{p^e}$. Since $(A/N)^{p^i} \cap (A/N)^{p^e} = (A/N)^{p^i} = ((A/N)^{p^e})^{p^{i-e}}$ for $i = e, e+1, \ldots,$ it follows that $g(A^{p^i} \cap K) = (A/N)^{p^i} \cap K$, $i = 1,2,\ldots$, $K(= gK = (A/N)^{p^e})$. Clearly A/N over K has a modular base. Thus the conclusion follows from Proposition 3.13. q.e.d.

<u>3.15. Corollary</u>. Let A and R be commutative rings with identity such that $A \supseteq R$ and the identities of A and R coincide. Let N and M be maximal ideals of A and R, respectively, such that $M = R \cap N$. Suppose $A^{p^{e+1}} \subseteq R \subseteq A^{p^e}$ for some non-negative integer e, p = characteristic of R. If

(a) R is a complete local ring (not necessarily Noetherian), or

(b) $A^{p^{e+1}}$ is a complete local ring (not necessarily Noetherian),

then there exists a coefficient field of R which is extendable to one of A.

Proof. (a) Since $A/N^{p^{e+1}} \subseteq R/M \subseteq (A/N)^{p^e}$, $(A/N)^{p^e}$ has a modular base over R/M. Now clearly $g(A^{p^{e+i}} \cap R) = (A/N)^{p^{e+i}} \cap (R/M)$, $i = 1,2,\ldots$. Thus there exists a coefficient field of R which is extendable to one of A^{p^e}, say E, by Proposition 3.13. By identifying E and R in Corollary 3.14, E can be extended to a coefficient field of A.

(b) Since $(A/N)^{p^{e+1}} \subseteq R/M \subseteq (A/N)^{p^e}$, there exists a p-base $G \cup D$ of $(A/N)^{p^e}$ such that $(A/N)^{p^{e+1}}(G) = R/M$ and $(R/M)(D) = (A/N)^{p^e}$. Let $G' \subseteq R$ and $D' \subseteq A^{p^e}$ be sets of representatives of G and D, respectively. Now $G^p \cup D^p$ is a p-base of $(A/N)^{p^{e+1}}$ and $G'^p \cup D'^p \subseteq A^{p^{e+1}}$. Since $A^{p^{e+1}}$ is complete, there exists a coefficient field E of $A^{p^{e+1}}$ such that $E \supseteq G'^p \cup D'^p$. Hence $K = E[G']$ and $E' = K[D']$ are coefficient fields of R and A^{p^e}, respectively. By Corollary 3.14, E' can be extended to a coefficient of A.

q.e.d.

<u>3.16. Proposition</u>. Let A be a quasi-local ring with unique maximal ideal N and let R be a complete local ring (not necessarily Noetherian) of prime characteristic p such that $A \supseteq R$, $M = R \cap N$ and the identities of A and R coincide. If $N^{p^e} = (0)$ for some positive integer e and A/N is separable over R/M, then every coefficient field of R is extendable to one A.

Proof. Let K be any coefficient field of R. Since A/N is separable over R/M, every p-base of R/M is extendable to one of A/N. Hence let $B \cup D$ be a p-base of A/N where B is a p-base of R/M. Let $B' \subseteq K$ and $D' \subseteq A$ be sets of representatives of B and D, respectively. Since $N^{p^e} = (0)$, A^{p^e} is a field. Clearly $B'^{p^e} \cup D'^{p^e} \subseteq A^{p^e}$, whence $L = A^{p^e}[B', D']$ is a coefficient field of A as in the proof of Corollary 3.14. Now $L = A^{p^e}[B',D'] \supseteq K^{p^e}[B']$ $= K$. q.e.d.

In Proposition 3.8 conditions, which include the case F/K is purely inseparable, are given for $L \otimes F$ to have a $1 \otimes F$ - coefficient field. We now determine criteria for the existence of a $L \otimes 1$ - coefficient field when F/K is purely inseparable.

3.17. Lemma. Suppose either L/K or F/K is algebraic. Then there exists sets B_L in L and B_F in F such that $B_L \cup B_F$ is a p-base of LF.

Proof. It follows by Zorn's lemma that there exists a set B_L^* in L and a set B_F^* in F which are p-independent in LF and such that $(LF)^p(B_L^*) = (LF)^p(L) = L(F^p)$ and $(LF)^p(B_F^*) = (LF)^p(F) = F(L^p)$. Thus $(LF)^p(B_L^*,B_F^*) = LF$. Therefore there exist subsets B_L of B_L^* and B_F of B_F^* such that $B_L \cup B_F$ is a p-base of LF. q.e.d.

We may choose $B_L = B_L^*$ and $B_F \subseteq B_F^*$ or $B_L \subseteq B_L^*$ and $B_F = B_F^*$ so that either B_L or B_F is a relative p-base of LF/F or LF/L, respectively.

3.18. Lemma. Suppose F/K has exponent e. Then $(L \otimes F)^{p^e}[B']$ is a coefficient field of $L \otimes F$ where B' is any set of a representatives of a p-base B of LF.

Proof. Since $F^{p^e} \subseteq K$, $(L \otimes F)^{p^e} \subseteq L \otimes 1$. Thus $(L \otimes F)^{p^e}$

is a field mapping naturally onto $(LF)^{p^e}$. We have the desired result by identifying $L \otimes F$ with A and $(L \otimes F)^{p^e}$ with R in the proof of Corollary 3.14. q.e.d.

3.19. Proposition. Suppose F/K has exponent e. If there exists an intermediate field F' of F/K such that $L \otimes F'$ is a coefficient field of $L \otimes F$, then for all subsets B_L of L and B_F of F' such that $B_L \cup B_F$ is a p-base of $LF'(= LF)$, we have that (1) $L = K(L^{p^e})(B_L)$, (2) $F' = K(B_F)$ and (3) $K \subseteq (L \otimes F)^{p^e}[B_L \otimes 1, 1 \otimes B_F]$. Conversely, if there exist subsets B_L of L and B_F of F such that (1') $B_L \cup B_F$ is a p-base of LF, (2') $L = K(L^{p^e})(B_L)$ and (3') $K \subseteq (L \otimes F)^{p^e}[B_L \otimes 1, 1 \otimes B_F]$, then there exists an intermediate field F' of F/K such that $L \otimes F'$ is a coefficient field of $L \otimes F$, namely $F' = K(B_F)$.

Proof. Suppose $L \otimes F'$ is a coefficient field. By Lemma 3.17, there exist subsets B_L of L and B_F of F' such that $B_L \cup B_F$ is a p-base of $LF' = LF$. For all such p-bases, $(L \otimes F)^{p^e}[B_L \otimes 1, 1 \otimes B_F]$ is a coefficient field of $L \otimes F$ by Lemma 3.18. Since $(L \otimes F)^{p^e}[B_L \otimes 1, 1 \otimes B_F] \subseteq L \otimes F'$, $(L \otimes F)^{p^e}[B_L \otimes 1, 1 \otimes B_F] = L \otimes F'$. Thus $K \subseteq (L \otimes F)^{p^e}[B_L \otimes 1, 1 \otimes B_F]$ whence $K(L^{p^e})(B_L) \otimes K(B_F) = (L \otimes F)^{p^e}[B_L \otimes 1, 1 \otimes B_F] = L \otimes F'$. Hence $L = K(L^{p^e})(B_L)$

and $F' = K(B_F)$. Conversely suppose (1'), (2') and (3') hold. Now $(L \otimes F)^{p^e}[B_L \otimes 1, 1 \otimes B_F]$ is a coefficient field of $L \otimes F$ containing K by (3'). Hence by (2')
$(L \otimes F)^{p^e}[B_L \otimes 1, 1 \otimes B_F] = K(L^{p^e})(B_L) \otimes K(B_F) = L \otimes K(B_F)$.
Let $F' = K(B_F)$. q.e.d.

3.20. Corollary. Suppose F/K has exponent 1. Then there exists an intermediate field F' of F/K such that $L \otimes F'$ is a coefficient field of $L \otimes F$.

Proof. Choose B_L^* in L and B_F in F so that $B_L^* \cup B_F$ is a p-base of LF and $(LF)^p(B_L^*) = L(F^p) = L$. Then $L \supseteq K(L^p)(B_L^*) \supseteq (LF)^p(B_L^*) = L$ whence $L = K(L^p)(B_L^*)$. Since $K \subseteq (LF)^p(B_L^*)$ and $F^p \subseteq K$, $K \subseteq (L \otimes F)^p(B_L^* \otimes 1)$. q.e.d.

3.21. Corollary. Suppose F/K has exponent e. If there exist subsets B_L in L and B_F in F such that $B_L \cup B_F$ is a p-base of LF and $(LF)^{p^e}(B_L, C_F) = L$ where $C_F = \{b^{p^i} \mid b \in B_F,\ i$ is the exponent of b over $K\}$, then there exists an intermediate field F' of F/K such that $L \otimes F'$ is a coefficient field of $L \otimes F$.

Proof. We have $L \subseteq (LF)^{p^e}(B_L, C_F) \subseteq K(L^{p^e})(B_L) \subseteq L$. Hence $L = K(L^{p^e})(B_L)$ and $K \subseteq (L \otimes F)^{p^e}[B_L \otimes 1, 1 \otimes C_F] \subseteq (L \otimes F)^{p^e}[B_L \otimes 1, 1 \otimes B_F]$. (If it is assumed instead that

$K \subseteq (LM)^{p^e}(B_L, C_F)$, then $L \otimes F$ has a K-coefficient field.)

q.e.d.

3.22. Example. Let $K = P(u,v)$ where P is a perfect field and u, v are independent indeterminates over P. Let $L = K(c^{p^{-f}})$ and $F = K(u^{p^{-e}}, v^{p^{-e}})$ where c is a root of the separable irreducible polynomial $x^2 + ux + v$ (in an indeterminate x) over K and e, f are positive integers [43]. If $f \geq e$, then $B_L \cup B_F$ is a p-base of LF where $B_L = \{c^{p^{-f}}\}$ and $B_F = \{u^{p^{-e}}\}$. Clearly $B_L \cup C_F$ is a p-base of L where $C_F = \{u\}$. Thus by Corollary 3.21 $L \otimes K(u^{p^{-e}})$ is a coefficient field of $L \otimes F$. If $f \leq e$, then $K(c) \otimes F$ is a coefficient field of $L \otimes F$ by Proposition 3.8.

3. Field composites. When L/K is purely inseparable and F/K is an arbitrary extension field, the analysis of the radical of $L \otimes F$ is useful for determining the structure of the unique field composite LF. We now extend to fields which have an exponent the basic methods and some of the results of the finite degree case [43]. L/K will be an extension with exponent throughout this section.

In the statement of the following proposition we use the implifying fact that $F(LF)^{p^j} = FL^{p^j}$. Also we omit the trivial cases where one or both exponents are zero.

3.23. Proposition. Let L/K and LF/F have positive exponents e and e', respectively. Then there exist two chains of subsets of L

$$\begin{aligned} &M_o \supseteq \cdots \supseteq M_i \supseteq \cdots \supseteq M_{e-1} \supseteq M_e = \emptyset \\ &M'_o \supseteq \cdots \supseteq M'_j \supseteq \cdots \supseteq M'_{e'-1} \supset M'_{e'} = \emptyset \end{aligned} \tag{1}$$

such that

(a) $M_i^{p^i}$ and $M_j'^{p^j}$ are minimal generating sets for $K(L^{p^i})/K$ and $F(L^{p^j})/F$, respectively $(i = 0,\ldots,e-1;\ j = 0,\ldots,e'-1)$, and

(b) for each j there exists at least one j' such that

$$M_{j'} \supseteq M'_j \supseteq M_{j'+1} \text{ and } j' \geq j. \tag{2}$$

Proof. Consider the proposition which we shall denote by $A(s)$, where s is an integer such that $1 \leq s \leq e$: There exists an integer s^*, $0 \leq s^* \leq e'$, and two chains of subsets of L, namely,

$$\begin{aligned} &M_{e-s} \supseteq \cdots \supseteq M_i \supseteq \cdots \supseteq M_{e-1} \supseteq M_e \\ &M'_{e'-s^*} \supseteq \cdots \supseteq M'_j \supseteq \cdots \supseteq M'_{e'-1} \supseteq M'_{e'} \end{aligned}$$

such that:

(1) The sets M_i and M'_j $(i = e-s,\ldots,e-1;$ $j = e'-s^*,\ldots,e'-1)$ have the properties (a) and (b) of Proposition 3.23.

(2) Either $s^* = e'$ (hence $M'_{e'-s^*} = M'_o$), or $s^* < e'$ and

(i) $M_{e-s}^{p^{e'-(s^*+1)}}$ is relatively p-independent in $F(L^{p^{e'-(s^*+1)}})/F$ and

(ii) if $M'_{e'-s^*} = M_{e-s}$ then $M'_{e'-s^*+1} \neq M_{e-s}$.

First we note that $A(e)$ is equivalent to the statement of Proposition 3.23. For, when $s = e$, then either $e^* = e'$ and we are done, or $e^* < e'$ and $M_o^{p^{e'-(e^*+1)}}$ is relatively p-independent in $F(L^{p^{e'-(e^*+1)}})/F$. In the latter case, M_o is relatively p-independent in $F(L)/F$. Since $LF = F(K(M_o)) = F(M_o)$, we may therefore set $M'_i = M_o$ for $i < e' - e^*$ to complete the second chain.

To prove $A(1)$ we consider the set

$$S_1^{(1)} = \{I \mid I \subseteq L,\ I^{p^{e-1}} \text{ is relatively p-independent in } K(L^{p^{e-1}})/K,\ I^{p^{e'-1}} \text{ is relatively p-independent in } F(L^{p^{e'-1}})/F\}.$$

Since $K \nsupseteq K(L^{p^{e-1}})$ and $F \nsupseteq F(L^{p^{e'-1}})$, there exist elements $a_1, a_2 \in L$ such that $a_1^{p^{e-1}} \notin K$ and $a_2^{p^{e'-1}} \notin F$. If

$a_1^{p^{e'-1}} \notin F$ then $\{a_1\} \in S_1^{(1)}$, or if $a_2^{p^{e-1}} \notin K$ then $\{a_2\} \in S_1^{(1)}$. But if neither of these negations hold, then $(a_1 + a_2)^{p^{e-1}} \notin K$ and $(a_1 + a_2)^{p^{e'-1}} \notin F$, which implies $\{a_1 + a_2\} \in S_1^{(1)}$. Thus $S_1^{(1)} \neq \emptyset$, and from the elementary properties of relative p-independence it follows that $S_1^{(1)}$ is inductive. Hence there exists a non-empty maximal set in $S_1^{(1)}$, say I_1. Suppose that $I_1^{p^{e-1}}$ is not a relative p-base for $K(L^{p^{e-1}})/K$ and that $I_1^{p^{e'-1}}$ is not a relative p-base for $F(L^{p^{e'-1}})/F$. Then there exist elements $a_1, a_2 \in L$ such that

$$a_1^{p^{e-1}} \notin K(I_1^{p^{e-1}}) \ [= K(L^{p^e})(I_1^{p^{e-1}})]$$

and

$$a_2^{p^{e'-1}} \notin F(I_1^{p^{e'-1}}) \ [= F(L^{p^{e'}})(I_1^{p^{e'-1}})].$$

If $a_1^{p^{e'-1}} \notin F(L^{p^{e'}})(I_1^{p^{e'-1}})$ then $I_1 \cup \{a_1\} \in S_1^{(1)}$, and if $a_2^{p^{e-1}} \notin K(L^{p^e})(I_1^{p^{e-1}})$ then $I_1 \cup \{a_2\} \in S_1^{(1)}$; each of these contradicts the maximality of I_1. But now we get $I_1 \cup \{a_1 + a_2\} \in S_1^{(1)}$, which is again a contradiction. Hence either I_1 can be chosen for M_{e-1} or if not then I_1 can be chosen for $M'_{e'-1} = \ldots = M'_{e'-t}$ for some maximal integer t, $1 \leq t \leq e'$. If in the former case the set I_1 can be chosen for $M'_{e'-1}$, then we do so and set $s^* = 1$, otherwise we let

$s^* = 0$. In the latter case, suppose $t = e'$. Then let $s^* = e'$ and construct a chain for L/K using I_1, namely $I_1 = M'_0 \subsetneqq M_{e-1} \subseteq \cdots \subseteq M_0$. But if $t < e'$, then we consider the set

$$S_1^{(2)} = \{I \mid M'_{e'-1} \subseteq I \subseteq L, \ I^{p^{e-1}} \text{ is relatively p-independent in } K(L^{p^{e-1}})/K, \ I^{p^{e'-(t+1)}} \text{ is relatively p-independent in } F(L^{p^{e'-(t+1)}})/F\}.$$

Let I_2 be a maximal element of $S_1^{(2)}$. If $I_2^{p^{e-1}}$ is not a relative p-base for $K(L^{p^{e-1}})/K$ and $I_2^{p^{e'-(t+1)}}$ is not a relative p-base for $F(L^{p^{e'-(t+1)}})/F$, then we obtain a contradiction of the maximality of I_2 just as in the case of I_1 above. Hence either I_2 can be chosen as M_{e-1} (and if I_2 can also be chosen as $M'_{e'-(t+1)}$ then we set $s^* = t+1$, otherwise $s^* = t$) or if not, then I_2 can be chosen as $M'_{e'-(t+1)}$ at least. In the latter case we proceed as in the case of I_1: Define $S_1^{(3)}$ (if $t+1 < e'$) and obtain I_3 and so forth, until either M_{e-1} is attained or, if M'_0 comes first, then construct M_{e-1} as an extension of M'_0: $M'_0 \subsetneqq M_{e-1} \subseteq \cdots \subseteq M_0$. Thus $A(1)$ is proved.

Now assume $A(s)$ is true and consider $A(s+1)$. If $M_{e-(s+1)}$ can be chosen as M_{e-s}, then $A(s+1)$ is already true with $(s+1)^* = s^*$. If not, then suppose $M'_{e'-s^*} = \cdots = M'_{e'-(s^*+t)}$ for t maximal. Consider the set

$S_{s+1}^{(1)} = \{I \mid M_{e-s} \subseteq I \subseteq L,\ I^{p^{e-(s+1)}}$ is relatively p-independent in $K(L^{p^{e-(s+1)}})/K$, $I^{p^{e'-(s^*+t+1)}}$ is relatively p-independent in $F(L^{p^{e'-(s^*+t+1)}})/F\}$.

A maximal element of $S_{s+1}^{(1)}$ is either an $M_{e-(s+1)}$ or an $M'_{e'-(s^*+t+1)}$. Proceeding as in the case A(1), we obtain an $M_{e-(s+1)}$ and an $M'_{e'-(s+1)^*}$ ($\subseteq M_{e-(s+1)}$). Finally, we note that $M_j'^{p^n}$ is relatively p-independent in $K(L^{p^n})/K$ for $n \leq j$ because $M_j' \subseteq L$ and $M_j'^{p^n}$ is relatively p-independent in $F(L^{p^n})/F$ for $n \leq j$. Hence if j' is the largest non-negative integer such that $M_j'^{p^{j'}}$ is relatively p-independent in $K(L^{p^{j'}})/K$, then $M_{j'+1} \subseteq M_j' \subseteq M_{j'}$ with $j' \geq j$. q.e.d.

3.24. Definition. The two chains (1) in Proposition 3.23 will be called compatible generating chains. We associate with such a pair of chains a pair of compatible canonical generating systems: $\{B_1,\ldots,B_e\}$ and $\{B_1',\ldots,B_{e'}'\}$, where $B_i = M_{i-1} - M_i$ and $B_j' = M_{j-1}' - M_j'$.

3.25. Remark. If L/K is an m-fold purely inseparable finite extension, then the original canonical description of this extension [43] has the form $L = K(a_1,\ldots,a_m)$ where

(a) $a_i^{q_i} \in K(a_1^{q_i},\ldots,a_{i-1}^{q_i})$, $q_i = p^{e_i}$, $e_i > 0$,

(b) $a_i^{p^{-1}} \notin K(a_1,\ldots,a_{i-1})$,

(c) $e_1 \geq e_2 \geq \ldots \geq e_m$.

When the finite degree case is thus expressed, Proposition 3.23 tells us that there exists an integer m', $m' \le m$, such that $LF/F = F(a_1,\ldots,a_{m'})$ where $a_1,\ldots,a_{m'}$ satisfy conditions analogous to (a), (b), (c). This fact is critical to the analysis of LF in [43]. Note that the ordering of the generators is the reverse of the ordering that arises naturally in the infinite degree case.

To effect a reduction from the case where L/K has an exponent to the case where L/K is of finite degree, the following proposition will be used.

3.26. Proposition. Let $M_0 \supseteq \ldots \supseteq M_{e-1}$ and $M'_0 \supseteq \ldots \supseteq M'_{e'-1}$ be compatible generating chains for L/K and LF/F, respectively. Let D be a finite subset of M_0. Then there exists a finite subset B^* of M_0 such that, with $L^* = K(B^*)$, we have

(a) $D \subseteq B^*$, $L^*F = F(B^* \cap M'_0)$,

(b) the finite degree extensions L^*/K and L^*F/F are canonically generated by B^* and $B^* \cap M'_0$ respectively, where the ordering of the generators is that induced by M_0 and where the structure of the canonical minimal polynomials of the generators is exactly the same in L^*/K and L^*F/F as in L/K and LF/F.

Proof. Let $\{B_1,\ldots,B_e\}$ and $\{B'_1,\ldots,B'_{e'}\}$ be the compatible canonical generating systems determined by the given chains. Apply Proposition 3.23 as follows: There exist maximal integers $i_2, \ldots, i_{e'}$ such that $M_{i_j-1} \supseteq M'_{j-1}$ and $i_j \geq j$ $(j = 2, \ldots, e')$, and

$$M_1 \supseteq \cdots \supseteq M_{i_2-1} \supseteq M'_1 \supseteq M_{i_2} \supseteq \cdots \supseteq M_{i_3-1} \supseteq M'_2 \supseteq M_{i_3} \supseteq \cdots$$
$$\supseteq M_{e'-1} \supseteq \cdots \supseteq M_{i_{e'}-1} \supseteq M'_{e'-1}.$$

Now, set $D = D_1$ and recall that

$$M_{i-1} = \bigcup_{n=i}^{e} B_n, \quad i = 1, \ldots, e,$$
$$M'_{j-1} = \bigcup_{n=j}^{e'} B'_n, \quad j = 1, \ldots, e'.$$

We commence the construction of B^* by considering $D_1 \cap B_1$. There exist finite (possibly empty) sets $D_2, \ldots, D_{i_2}$ such that $D_2 \subseteq M_1, \ldots, D_{i_2} \subseteq M_{i_2-1}$ and $(D_1 \cap B_1)^p \subseteq K(D_2^p), \ldots, [(D_1 \cup D_2 \cup \cdots \cup D_{i_2-1}) \cap B_{i_2-1}]^{p^{i_2-1}} \subseteq K(D_{i_2}^{p^{i_2-1}})$. Set $D_1^{(2)} = D_1 \cup D_2 \cup \cdots \cup D_{i_2}$. Consider $D_1^{(2)} \cap B'_1$. There exists a finite set $D'_2 \subseteq M'_1$ such that $(D_1^{(2)} \cap B'_i)^p \subseteq F(D_2'^p)$. Let $D_{12} = D_1^{(2)} \cup D'_2$. Since $M_{i_2-1} \supseteq M'_1 \supseteq D'_2$, we have $D_{12} \cap B_1 = D_1 \cap B_1, \ldots, D_{12} \cap B_{i_2-1} = (D_1 \cup D_2 \cup \cdots \cup D_{i_2-1}) \cap B_{i_2-1}$. Hence $(D_{12} \cap B_1)^p \subseteq K(D_2^p), \ldots, (D_{12} \cap B_{i_2-1})^{p^{i_2-1}} \subseteq K(D_{i_2}^{p^{i_2-1}})$. We also have $(D_{12} \cap B_1')^p = (D_1^{(2)} \cap B_i')^p \subseteq F(D_2'^p)$. Continuing, there exist finite sets $D_{i_2+1},\ldots,D_{i_3}$ such that

$D_{i_2+1} \subseteq M_{i_2},\ldots,D_{i_3} \subseteq M_{i_3-1}$ and $(D_{12} \cap B_{i_2})^{p^{i_2}} \subseteq K(D^{p^{i_2}}_{i_2+1}),\ldots,$

$[(D_{12} \cup D_{i_2+1} \cup \ldots \cup D_{i_3-1}) \cap B_{i_3-1}]^{p^{i_3-1}} \subseteq K(D^{p^{i_3-1}}_{i_3})$. Set

$D_{12}^{(3)} = D_{12} \cup D_{i_2+1} \cup \ldots \cup D_{i_3}$. There exists a finite set

$D_3' \subseteq M_2'$ such that $(D_{12}^{(3)} \cap B_2')^{p^2} \subseteq F(D_3')^{p^2}$. Set $D_{13} = D_{12}^{(3)} \cup D_3'$.

Since $M_1' \supseteq M_{i_2} \supseteq \ldots \supseteq M_{i_3-1}$ we have $D_{13} \cap B_1' = D_{12} \cap B_1'$.

Hence $(D_{13} \cap B_1')^p \subseteq F(D_2'^p)$. Since $M_{i_3-1} \supseteq M_2' \supseteq D_3'$, we have

$D_{13} \cap B_1 = D_{12} \cap B_1,\ldots,D_{13} \cap B_{i_2-1} = D_{12} \cap B_{i_2-1}$ and

$D_{13} \cap B_{i_2} = D_{12} \cap B_{i_2},\ldots,D_{13} \cap B_{i_3-1} =$

$(D_{12} \cup D_{i_2+1} \cup \ldots \cup D_{i_3-1}) \cap B_{i_3-1}$. Hence

$(D_{13} \cap B_1)^p \subseteq K(D_2^p),\ldots,(D_{13} \cap B_{i_2-1})^{p^{i_2-1}} \subseteq K(D^{p^{i_2-1}}_{i_2})$ and

$(D_{13} \cap B_{i_2})^{p^{i_2}} \subseteq K(D^{p^{i_2}}_{i_2+1}),\ldots,(D_{13} \cap B_{i_3-1})^{p^{i_3-1}} \subseteq K(D^{p^{i_2-1}}_{i_3})$.

Also, $(D_{13} \cap B_2')^{p^2} = (D_{12}^3 \cap B_2')^{p^2} \subseteq F(D_3'^{p^2})$. A finite number of steps yields $D_{1e'}$. A similar process for $M_{i_{e'}}$ with $D'_{e'+1} = \ldots = D'_e = \emptyset$, yields D_{1e}. Now $(D_{1e} - D_{1e'})^{p^{e'}} \subseteq F$ because $B'_{e'} \supseteq M_{i_{e'}}$. Set $B^* = D_{1e}$.

Then $L^*F = F(B^*) = F(B^* \cap M_o')$. If $b \in B^* - M_o'$, then b has canonical exponent 0 in LF/F and by the construction of B^*, b has canonical exponent 0 in L^*F/F. The conservation of the minimal polynomials follows from the fact that $(B^* \cap B_j)^{p^j} \subseteq K(B_{j+1}^{p^j}, \ldots, B_e^{p^j})$. q.e.d.

For a given pair of compatible generating chains, the construction of B^* from D will lead to a unique B^* if we require that D_k and D_k' be minimal. The latter choices are possible because $K(M_i^{p^i})/K$ is minimally generated by $M_i^{p^i}$, and similarly for $F(M_i'^{p^i})/F$.

Set $B_o' = M_o - M_o'$ and call $B_i \cap B_j'$ the i,j-cell where $i = 1,\ldots,e$; $j = 0,1,\ldots,e'$. Consider the cells as objects ordered lexicographically by the lexicographical ordering of the pairs (i,j). Each cell as a set is considered to have a fixed but arbitrary well-ordering imposed on it.

3.27. Definition. When the cells and their elements are ordered as described above, we say that the minimal generating sets M_o and M_o' are compatibly ordered.

If $b_\alpha \in B_i \cap B_j'$ and $b_\beta \in B_k \cap B_\ell'$ and $(i,j) < (k,\ell)$, then we sometimes write $b_\alpha < b_\beta$, $B_i \cap B_j' < B_k \cap B_\ell'$; for additional clarity we also use the notation $b_{\alpha,i,j}$ for b_α. Suppose M_o, M_o' are compatibly ordered and let $b_\alpha \in M_o$. In Proposition 3.26 let us set $D = \{b_\alpha\}$ and denote the uniquely associated B^* by $\{b_\alpha\}^*$. Then we spell out the latter set by

$$\{b_\alpha\}^* = \{b_\alpha, b_{\alpha(1)}, \ldots, b_{\alpha(n)}\},$$

where n depends on b_α and the elements are listed according to the compatible ordering. Thus $L^* = K(\{b_\alpha\}^*)$ and $L^*F = K(\{b_\alpha\}^* \cap M_0')$. The canonical degree of an element of B_0' is defined as 1 and is denoted by q_0'. We shall often abbreviate the canonical degree by using the notation $q(b)$ when $b \in M_0$ and $q'(b)$ when $b \in M_0'$. Thus in particular $q'(b) = 1$ when $b \in B_0'$.

Now, in L^*/K as well as in L/K we have the unique expression

$$b_\alpha^{q(b_\alpha)} = f(b_{\alpha(1)}^{q(b_\alpha)}, \ldots, b_{\alpha(n)}^{q(b_\alpha)}), \tag{3}$$

where $f(x_1, \ldots, x_n) \in K[x_1, \ldots, x_n]$ and $\deg x_k < q(b_{\alpha(k)})/q(b_\alpha)$. Similarly, in L^*F/F as well as in LF/F, we have the unique expression

$$b_\alpha^{q'(b_\alpha)} = g(b_{\alpha(1)}^{q'(b_\alpha)}, \ldots, b_{\alpha(n)}^{q'(b_\alpha)}), \tag{4}$$

where $g(x_1, \ldots, x_n) \in F[x_1, \ldots, x_n]$ and $\deg x_k < q'(b_{\alpha(k)})/q'(b_\alpha)$. Of course, f and g depend on b_α.

Let φ denote the canonical F-epimorphism of $L \otimes F$ onto LF. We denote the radical $(\mathrm{Ker}\ \varphi)$ of $L \otimes F$ by N. Set $C_i = B_i \otimes 1$, $i = 1, \ldots, e$; $C_i' = B_i' \otimes 1$, $i = 0, 1, \ldots, e'$; $C = M \otimes 1$, and $C' = M_0' \otimes 1$. Order these sets by the induced order. For clarity we also use the notation $c_{\alpha,i,j}$ for c_α. Finally, set $q(c_\alpha) = q(b_\alpha)$ and $r(c_\alpha) = q(c_\alpha)/q'(c_\alpha)$.

Clearly, (3) is satisfied in $L^* \otimes F$ as well as in $L \otimes F$, after replacing b by $c = \varphi(b)$.

Let F_c denote the following complete residue system for $L \otimes F/N$ as an algebra over F (that is, over $1 \otimes F$):

$$F_c = \{\sum F \prod c^{i(c)} \mid c \in C', \ 0 \leq i(c) < q(c)\}.$$

We use F_c to form our unique counterimages with respect to φ.

Now suppose for some b_α that $q(b_\alpha) = q'(b_\alpha)$, that is, $r(c_\alpha) = 1$. Then, with $c_{\alpha(n)}$ terms omitted for brevity,

$$f(c_{\alpha(1)}^{q'(c_\alpha)} \ldots) - g(c_{\alpha(1)}^{q'(c_\alpha)} \ldots) = 0.$$

When $r(c_\alpha) > 1$, we define w_α by (following Pickert)

$$w_\alpha = c_\alpha^{q'(c_\alpha)} - g(c_{\alpha(1)}^{q'(c_\alpha)}, \ldots) \tag{5}$$

and set $r(w_\alpha) = r(c_\alpha)$. Hence

$$w_\alpha^{r(w_\alpha)} = f(c_{\alpha(1)}^{q(c_\alpha)}, \ldots) - g^*(c_{\alpha(1)}^{q(c_\alpha)}, \ldots), \tag{6}$$

where g^* indicates that the coefficients of g are raised to the $r(c_\alpha)$ power. We note in particular, that if $b_\alpha \in M'_{e'-1}(= B_e \cap B'_{e'})$, then $w_\alpha^{r(w_\alpha)} = w_\alpha^{p^e} \in 1 \otimes F$. Since $\varphi(w_\alpha) = 0$, $w_\alpha^{r(w_\alpha)} = 0$ in this case.

Denote by W the set of elements w_α defined by (5), and order them as induced by our original compatible ordering.

3.28. Proposition. The following set S of distinct power products is a linear basis of N (over F):

$$S = \{\Pi\, c^{i(c)} w^{j(w)} \mid c \in C', w \in W, 0 \leq i(c) < q'(c), 0 \leq j(w) < r(w), \sum j(w) > 0\}.$$

Proof. Let S' be a finite subset of S. Each element of S' involves a finite subset of C' and a finite subset of W. Since the defining equations for the relevant elements of W involve only a finite subset of C', the union T of all the finite subsets of C' thus involved is itself finite. Set $D = \varphi(T)$ in Proposition 3.26 to obtain L^* and $L^* \otimes F$. Then S' is a subset of $L^* \otimes F$, so by the known finite degree case [43, p.100] we find that S' is linearly independent over F. Finally we note that $\operatorname{Ker} \varphi = FS$, so S generates N over F.
q.e.d.

Since there exist only a finite number of cells, the following lemma is readily deduced from the finite degree case [42, p. 99]. Recall that $r(c_\alpha) > 1$ if and only if $w_\alpha \neq 0$, and that by (5):

$$c_\alpha^{q'(c_\alpha)} = w_\alpha + g(c_{\alpha(1)}^{q'(c_\alpha)}, \ldots)$$

Hence, whenever a power of c_α occurs with an exponent $\geq q'(c_\alpha)$ in a power product $\Pi\, c^{i(c)}$, where $c \in C'$ and

$0 \leq i(c) < q(c)$, then this power of c_α can be replaced by a sum involving only (a) a power of c_α with exponent $< q'(c_\alpha)$, (b) w_α, and (c) certain c_β's with $c_\alpha < c_\beta$.

3.29. Lemma. A. Every power product in a finite number of elements of C' belonging to cells $\geq B_i \cap B_j'$ for a fixed (i,j), can be expressed as a polynomial in w's from cells $\geq B_i \cap B_j'$ with coefficients in F_c involving only c's from cells $\geq B_i \cap B_j'$. B. If (i,j) is fixed and $w_\alpha = w_{\alpha,i,j}$, then

$$w_\alpha^{r(w_\alpha)} = F_\alpha(w_{\alpha(1)},\ldots,w_{\alpha(m)}) \tag{8}$$

where $w_\alpha < w_{\sigma(s)}$, $w_{\sigma(s)}$ occurs with an exponent $< r(w_{\sigma(s)})$, $s = 1,\ldots,m$, coefficients of F_α are in F_c and involve only c's from cells $> B_i \cap B_j'$. C. Every polynomial in w's which belong to cells $\geq B_i \cap B_j'$ and has coefficients from F_c, can be expressed as a polynomial in w's from the same set of cells but of degree $< r(w)$ in each w and with coefficients from F_c involving (in addition to initially occurring elements of C') only c's which belong to cells $\geq B_i \cap B_j'$.

3.30. Definition. A polynomial expression in elements of F_c and W of the final form described in part C. of the Lemma 3.29, is called reduced. In particular, every element of F_c is reduced.

It is readily shown that the reduced polynomial F_α of (8) above has no term free of w's, which is indeed a necessary and sufficient condition for a reduced polynomial to be an element of N.

Let i be a positive integer. In the remainder of these notes, let N^i denote the usual sum of products of i elements of N, and let N_i be defined by

$$N_i = \{\Sigma F_c \Pi w^{i(w)} \mid w \in W, 0 \leq i(w) < r(w), \Sigma i(w) \geq i\}.$$

Note that N^i and N_i are F-modules. Finally, let $\overline{W}_i$ denote the set of cosets defined by

$$\overline{W}_i = \{\Pi w^{i(w)} + N^{i+1} \mid w \in W, 0 \leq i(w) < r(w), \Sigma i(w) = i\},$$

with $\overline{W}_o = \overline{W}$. The following is readily verified.

3.31. Remark. Let i be a fixed positive integer. $N^i = N_i$ and $N^{i+1} = N_{i+1}$ if and only if $\overline{W}_i$ is a linear basis of N^i/N^{i+1} as a LF-module. In particular $N^2 = N_2$ if and only if $\overline{W}$ is a linear basis of N/N^2 (that is, a minimal ideal basis of N).

3.32. Proposition. If those $q'(b_\alpha)$'s for which $q'(b_\alpha) < q(b_\alpha)$ are all equal, then

$$N^i = N_i \quad \text{for all } i. \tag{9}$$

Proof. By Proposition 3.26, w_α is an element of some $L^* \otimes F$ in which it maintains the relation (8). Since L^*/K is of finite degree, we can apply the known finite degree result [43, p.101] to the radical $N(B^*_\alpha)$ of $L^* \otimes F$, where B^*_α denotes a minimal generating set for L^*/K as determined by Proposition 3.26. We have that $N(B^*_\alpha)^i = N(B^*_\alpha)_i$ for all i. Clearly $N = \cup \{N(B^*_\alpha) \mid \text{all } B^*_\alpha \text{ for all } w_\alpha\}$, $N^i = \cup N(B^*_\alpha)^i$ and $N_i = \cup N(B^*_\alpha)_i$. Hence $N^i = N_i$. q.e.d.

<u>3.33. Corollary</u>. Given the hypothesis of Proposition 3.32 we have

(a) $|W| = |B'_o|$ if $M'_o \neq M_o$,

(b) $|W| \leq |B'_j|$ for some j $(1 \leq j \leq e')$ if $M'_o = M_o$.

Proof. The q'_j's are distinct. Hence elements with equal q' must be in the same B'_j. Note that $q(b) \geq p$, all $b \in M_o$. q.e.d.

When $L \supseteq F \supseteq K$, then F/K is also purely inseparable of bounded exponent. Hence we can consider $F \otimes L$ over $1 \otimes L$ along with its image LF/L. Since $F \otimes L$ and $L \otimes F$ have identifiable radicals under the natural isomorphism between $L \otimes F$ and $F \otimes L$, and since (9) holds for $F \otimes L$ because every q' equals 1, (9) also holds for $L \otimes F$. Thus we have the following proposition concerning relative imperfection degrees:

3.34. Proposition. If $L \supseteq F \supseteq K$, then (relative imperfection degree of L/K) $\geq$ (relative imperfection degree of F/K) $= |W|$.

Proof. Reverse the roles of L and F and let M_o' and M_o'' be relative p-bases of F/K and LF/L, respectively, so $M_o'' = \emptyset$. Thus $M_o' - M_o'' = M_o'$, whence by Corollary 3.33 $|W| = |M_o'|$. Since $N^2 = N_2$ we have for $L \otimes F$ over F that $[N/N^2 : LF] = |W|$ and from our construction of W we have $|W| \leq |M_o|$, where M_o is a relative p-base of L/K. q.e.d.

Consider compatible canonical generating systems for L/K and LF/F derived from compatible generating chains. When $b \in B_i \cap B_j'$ $(j > 0)$, let us set

$$E(b) = e_i \quad \text{and} \quad E'(b) = e_j',$$

where e_i and e_j' are the canonical exponents corresponding to B_i and B_j', respectively. Call $E(b)$ and $E'(b)$ the cell exponents of b. The following theorem appears to be the natural extension of a known finite degree result [43, Satz 30].

3.35. Proposition. Suppose $L \supseteq F \supseteq K$ and F/K is a simple extension. Then for the cell exponents corresponding to a given pair of compatible generating systems, we have $E(b_\alpha) \leq E'(b_\beta) \leq E(b_\beta)$ whenever $b_\alpha < b_\beta$.

Proof. $E'(b_\beta) \le E(b_\beta)$ is always true. Now apply Proposition 3.26 to obtain $L^* \supseteq F \supseteq K$ such that the non-empty cells of L/K, LF/F are in $1-1$ correspondence with those of L^*/K, L^*F/F. Starting with greatest cell in the latter pair, Satz 30 in [43] implies that $E(b) = E'(b)$ for every b in a given cell and that $E'(b_\beta) \ge E(b_\alpha)$ when b_α is in the cell immediately less than that of b_β. q.e.d.

The proof of Satz 30 in [43, p.104] depends on the finite degree case of the following theorem, which is an extension of Satz 29 in [43] and deals with the defect m_1 of N, where $m_1 = [N/N^2 : LF]$. Suppose $|W| = n < \infty$, so we may write $W = \{w_1, \ldots, w_n\}$ temporarily for the set of distinct elements of W. Let $z_1, \ldots, z_n$ be independent indeterminates over $L \otimes F$. Now, for equation (8), we assume $w_1, \ldots, w_n$ are in canonical order. Then $w_i^{r_i} = F_i(w_{i+1}, \ldots, w_n)$. With the obvious notation $u_{jk}(c) \in F_c$, $\varphi(u_{jk}(c)) = u_{jk}(b)$ and $u_{jk}(c) = 0$ for $j \ge k = 1, \ldots, n$, we have

$$F_i(w_{i+1}, \ldots, w_n) = \sum_{k=1}^{n} u_{n-i+1,k}(c) w_{n-k+1} + \ldots, \quad i = 1, \ldots, n.$$

Hence

$$F_i(z_{i+1}, \ldots, z_n) = \sum_{k=1}^{n} u_{n-i+1,k}(c) z_{n-k+1} + \ldots, \quad i = 1, \ldots, n. \qquad (10)$$

The terms written out constitute the linear part of $F_i(z_{i+1}, \ldots, z_n)$.

<u>3.36. Proposition</u>. If $|W| = n < \infty$ and $F_i(z_{i+1},\ldots,z_n)$ is the polynomial in equation (10), then the matrix $(u_{jk}(b))$, $j = 2,\ldots,n$ and $k = 1,\ldots,n-1$, has rank $n - m_1$.

Proof. Barred letters indicate residue classes mod N^2. Since $\sum u_{n-i+1,k}(b)\bar{w}_{n-k+1} = \bar{0}$, the rows of the $n \times n$ matrix U (obtained by augmenting $(u_{jk}(b))$ with a zero first row and zero last column) display coefficients in linear combinations of the w_i's which equal 0 mod N^2. As in the proof of Satz 29, we must show that every linear dependence is a linear combination of the linear dependences which occur from the rows of U. That is, if $\sum v_i(b)\bar{w}_i = \bar{0}$, then $\vec{v} = \vec{h}U(\text{mod } N^2)$ for some n-tuple $\vec{h}$. We can write the latter premise as

$$\sum v_i(c)w_i = H(w_1,\ldots,w_n) \in N^2, \qquad (11)$$

where the left-hand side represents the reduction of the right-hand side. We now combine all the w's with all the c's that occur in (11) and all the c's required to express H as $\sum nn'$, where $n, n' \in N$. (Note we already have all the w's.) Altogether this yields a finite set D of c's which by application of Proposition 3.26 provides a finite extension L^*/K such that in $L^* \otimes F$ equation (11) holds and $H \in N^{*2}$ (N^* the radical of $L^* \otimes F$). That is, in $L^* \otimes F$

$$\sum v_i(c)w_i = H(w_1,\ldots,w_n) \in N^{*2}.$$

Now we may apply the argument used in the finite degree proof to conclude that in $L^* \otimes F$ we have $\vec{v} = \vec{h}U$. Since U is constructed independently of L^*, we get $m_1 = n - \text{rank } U$.

q.e.d

The effect on the canonical exponents of the adjunction of a p-th root is given by the following proposition (see [43], p.106). Here the canonical generating systems used are those described in p.32, so only non-empty B_i's are used.

<u>3.37. Proposition</u>. Suppose $L \supseteq F \supseteq K$, $\{B_1, \ldots, B_r\}$ is the canonical generating system of p.32 for F/K with canonical exponents $e_1 < \ldots < e_r$, and $L = F(a)$ with exponent 1. Then the set of canonical exponents of L/K is one of the following types, where h is a non-negative integer and $e_o = 0$:

(1) the same as for F/K,

(2) $e_1 < \ldots < e_{h-1} < e_{h+1} < \ldots < e_r$,

(3) $e_1 < \ldots < e_h < e_h + 1 < e_{h+1} < \ldots < e_r$,

(4) $e_1 < \ldots < e_{h-1} < e_h + 1 < e_{h+1} < \ldots < e_r$.

Proof. If $a^p \in K(F^p)$, then $\{a\} \cup B_1 \cup \ldots \cup B_r$ is a minimal generating set for L/K. By Proposition 1.37, we have type (1) if $e_1 = 1$ and type (3) with $h = 0$ if $e_1 > 1$. If $a^p \in F - K(F^p)$, then $\{a^p\}$ is in a minimal generating set B' of F/K. Let $B' = B_1' \cup \ldots \cup B_r'$, where $\{B_1', \ldots, B_r'\}$ is a canonical generating system of F/K and $a^p \in B_h'$. Then $(B' - \{a^p\}) \cup \{a\}$ is a minimal generating set for L/K for some $a' \in B'$. Using Proposition 1.37 and the invariance properties of canonical generating systems, we have the following possibilities: Since a has exponent $e_h + 1$ over $K' = K(B'_{h+1},\ldots,B'_r)$, $\{\{a\} \cup B'_{h+1},\ldots,B'_r\}$ is a canonical generating system of $K'(a)/K$ and $e_h + 1 = e_{h+1} < \ldots < e_r$, or $\{\{a\}, B'_{h+1},\ldots,B'_r\}$ is a canonical generating system of $K'(a)/K$ and $e_h + 1 < e_{h+1} < \ldots < e_r$. Since $L \cong K'(a) \otimes_{K'(a^p)} F$, a linear basis for $F/K'(a^p)$ is one for $L/K'(a)$. Thus $\{B'_1,\ldots,B'_{h-1},B'_h - \{a^p\}\}$ is a canonical generating system for $L/K'(a)$ if $B'_h \supset \{a^p\}$, or $\{B'_1,\ldots,B'_{h-1}\}$ is a canonical generating system for $L/K'(a)$ if $B'_h = \{a^p\}$. q.e.d.

Reference Notes for Chapter III

The connections between modularity and extension coefficient fields of tensor products were made in [32], [31], [38]. Some of the additional results in Part A concerning extension coefficient fields of tensor products and other commutative algebras are in [4], [39], [29], [30], [34]. The results on composites are mainly from [36], [10]. As the references to Pickert [43] make clear, the main purpose of Part B is to show how the finite degree assumption in [43] can be relaxed.

References

1. Becker, M. T. and S. MacLane, The minimum number of generators for inseparable algebraic extensions, Bull. Am. Math. Soc. 46(1940) 182-186.

2. Berger, R., Differentiale hoherer Ordnung und Korpererweiterungen bei Primzahl charakeristik, Sitz. der Heid. Ak. der Wiss. 3 Abhandlung 1966.

3. Bryant, S. J. and J. L. Zemmer, A note on completely primary rings, Proc. Am. Math. Soc., 8(1957) 1940-1.

4. Chang, C. L., J. N. Mordeson, B. Vinograde, Note on coefficient fields in tensor products, The Formosan Science 21(1967) 5-7.

5. Chase, S., Inseparable Galois Theory and a theorem of Jacobson, Preprint, 1969.

6. Chase, S. and M. Sweedler, Hopf algebras and Galois Theory. Springer-Verlag Lecture Notes in Mathematics, 87, 1969.

7. Cohen, I. S., On the structure and ideal theory of complete local rings, Trans. Am. Math. Soc. 59(1946) 54-106.

8. Davis, R. L., A Galois Theory for a class of purely inseparable exponent two field extensions, Bull. Am. Math. Soc. 75(1969) 1001-1004.

9. Dieudonne, J., Sur les extensions transcendantes separable, Sum. Bras. Math. vol.II(1947) 1-20.

10. Eke, B. I., Inseparable composites. Submitted for publication.

11. Gerstenhaber, M., On the Galois theory of inseparable extensions, Bull. Am. Math. Soc. 70(1964) 561-66.

12. __________, On infinite inseparable extensions of exponent one, Bull. Am. Math. Soc. 71(1965) 878-881.

13. __________, On modular field extensions, J. of Alg. 10 (1968), 478-484.

14. Haddix, G. and J. N. Mordeson, Subbases applied to the existence of coefficient fields in commutative rings, The Formosan Science 21(1967) 1-4.

15. Haddix, G., J. Mordeson and B. Vinograde, Lattice isomorphisms of purely inseparable extensions, Math. Z. III(1969) 169-174.

16. __________, On purely inseparable extensions of unbounded exponent, Can. J. Math. XXI(1969) 1526-1532.

17. __________, On the two main field towers of a purely inseparable extension. Submitted for publication.

18. Hamann, E. and J. Mordeson, Pure inseparable field extensions, Math. Z. 103(1968) 43-47.

19. Heerema, N., Derivations and embeddings of a field in its power series ring, Proc. Am. Math. Soc. 11(1960) 188-194.

20. __________, Derivations and embeddings of a field in its power series ring. II, Mich. Math. J. 8(1961) 129-134.

21. __________, Convergent higher derivations on local rings, T.A.M.S. 132(1968) 31-44.

22. __________, A Galois Theory for inseparable field extensions. Preprint, 1970.

23. Jacobson, N., Galois Theory of purely inseparable extensions of exponent one, Am. J. of Math. 46(1944) 878-881.

24. __________, Lectures in Abstract Algebra, Vol.III, Van Nostrand, Princeton, N.J., 1964.

25. Kraft, H., Inseparable Korpererweiterungen, Com. Math. Hel. 45(1970) 110-118.

26. MacLane, S., A lattice formulation for transcendence degrees and p-bases, Duke Math. J. Vol.4 (1938) 455-468.

27. __________, Subfields and automorphism groups of p-adic fields, Ann. of Math. 40(1939) 423-442.

28. __________, Modular fields I. Separating transcendence bases, Duke Math. J. 5(1939) 372-393.

29. Mordeson, J. N., Remark on coefficient fields in complete local rings, J. Math. Kyoto Univ. 4(1965) 637-639.

30. Mordeson, J. N. and B. Vinograde. Extensions of certain subfields to coefficient fields in commutative algebras, J. Math. Soc. Japan 17(1965) 47-51.

31. Mordeson, N. and B. Vinograde, Generators and tensor factors of purely inseparable fields, Math. Z. 107(1968) 326-334.

32. __________, Tensor products of purely inseparable field extensions, Proc. Am. Math. Soc. 20(1969) 209-212.

33. __________, Note on relative p-bases of purely inseparable extensions, Proc. Am. Math. Soc. 22(1969) 587-590.

34. __________, Remarks on coefficient fields of tensor products, The Formosan Science 23(1969) 1-3.

35. __________, Exponents and intermediate fields of purely inseparable extensions, J. of Alg. To appear.

36. __________, Field composites with a bounded purely inseparable factor, Math. Z. To appear.

7. __________, Exponents of purely inseparable extensions, Archiv f. Math. To appear.

8. __________, Note on modular field extensions. (Submitted for publication.)

9. Nagata, M., Note on coefficient fields of complete local rings, Mem. Coll. Sci. Univ. Kyoto 32(1959-60) 91-92.

0. Nagata, M., Local Rings, Interscience Pub., Vol.13, 1962.

1. Ojanguren, M. and R. Sridharan, A note on purely inseparable extensions, Com. Math. Hel. 44(1969) 457-461.

2. Pickert, G., Neue Methoden i.d. Strukturtheorie d. komm. assoz. Algebren, Math. Ann. 116(1939) 217-280.

3. __________, Inseparable Korpererweiterungen, Math. Z. 52(1949) 81-135.

4. __________, Eine Normal form fur endliche rein-inseparable Korpererweiterungen, Math. Z. 53(1950), 133-135.

5. __________, Zwischenkorperverbande endlicher inseparabler Erweiterungen, Math. Z. 55(1952) 355-363.

6. Ponomarenko, P., The Galois Theory of infinite purely inseparable extensions, Bull. Am. Math. Soc. 71(1965), 878-881.

47. Reid, J. D., A note on inseparability, Mich. Math. J. 13(1966) 219-223.

48. Rygg, P., On minimal sets of generators of purely inseparable field extensions, Proc. Am. Math. Soc. 14(1963), 742-745.

49. Rygg, P. and B. Lehman, A note on an equivalence relation on a purely inseparable extension, Can. Math. Bull. 12 (1969) 175-178.

50. Schmidt, F. K. and S. MacLane, The generation of inseparable fields, Proc. N.A.S. 27(1941) 583-587.

51. Seligman, G. B., Modular Lie Algebras. Ergeb. d. Math. 40, Springer-Verlag, 1967.

52. Shatz, S., Galois Theory, Proc. Battelle Conference on Categorical Algebra, Springer-Verlag Lecture Notes in Mathematics , 1969.

53. Sweedler, M., Structure of inseparable extensions, Ann. of Math. 87(1968) 401-410.

54. __________, (Correction to above paper), Ann. of Math. 89(1969) 206-207.

55. __________, The Hopf algebra of an algebra applied to field theory, J. of Alg. 8(1968) 262-276.

56. Teichmuller, O., p-algebren, Deutsche Math. 1(1936) 362-368.

57. Weisfeld, M., Purely inseparable extensions and higher derivations, Trans. Am. Math. Soc. 116(1965) 435-469.

58. Zariski, O. and P. Samuel, Commutative Algebra, Vol.2 (Van Nostrand, Princeton, N. J., 1960).